The Production and Application
of Fluorescent Brightening Agents

The Production and Application of Fluorescent Brightening Agents

MILOŠ ZAHRADNÍK

Translation Želimír Procházka

A Wiley-Interscience Publication

JOHN WILEY & SONS
Chichester • New York • Brisbane • Toronto • Singapore

John Wiley & Sons Ltd., Chichester, Sussex
in co-edition with SNTL, Publishers of Technical Literature, Prague.

Library of Congress Cataloging in Publication Data:

Zahradník, Miloš.
   The production and application of fluorescent
   brightening agents.

   /A Wiley-Interscience publication./
   Includes index.
   1. Optical brighteners.   I. Title.
   TP894.5.Z3313   667   81-16325
   ISBN 0 471 10125 7   AACR2

British Library Cataloging in Publication Data:

Zahradník, Miloš
   The production and application of fluorescent
   brightening agents.
   1. Optical brighteners
   I. title
   667'.2   TP894.5

   ISBN 0 471 10125 7

Printed in Czechoslovakia

Contents

1
Introduction

The commonly held belief during the past half century that dye chemistry had reached its ultimate stage of development has been disproved at least twice in recent years: firstly by the invention of reactive dyes and secondly by the rapid development of optical brightening agents. As far as the latter compounds are concerned, there does not exist any textbook devoted exclusively to these, although several review articles have appeared. This fact gave the first impulse for the writing of this book. The second factor which inspired an interest in publication was the major contribution made by this narrow group of "colourless dyes" to so many branches of industry, and the wide range of possible applications of dyes with fluorescent, phototropic and thermotropic properties, on which all the interest of further research and development in the chemistry of dyes will evidently focus. The production of optical brightening agents does not closely parallel the technology of the production of dyes and it uses as raw materials and semi-finished products substances which usually do not belong to the production requirements of dyeing products. In addition to this (especially with respect to the intermediates used for disperse optical brighteners) optical brighteners are often very complex and required in a highly pure state. These differences in technological requirements presented a further stimulus for the publication of more detailed information on the problems of the production and development of optical brighteners, which would serve those who produce, and those who apply them.

In this book the most important technical data on the production and use of optical brightening agents are summarized in a rather detailed form, so that a wide circle of people can find the necessary information in it. The introductory chapters discuss both the fundamental theoretical principles of absorption and luminiscence, as well as clear examples of the technology of the production of some actual brands of optical brighteners.

1.1 Historical development of optical brightening agents

Numerous materials, especially textiles, both classical (cotton, wool, linen, silk) and synthetic (mainly polyamide, polyester and polyacrylonitrile), are not completely white and efforts have been made since ancient times to free them from their yellowish tinge or even locally stronger colouration. Bleaching in the sun, blueing and later chemical bleaching of textile and other materials increased the brightness of the products and eliminated to a certain extent the yellowish to greyish-yellow hue or the local impurity of the original or industrially treated material. In 1852 the outstanding physicist George Gabriel Stokes (1820 to 1903) elaborated the laws of fluorescence, named after him, which contributed the theoretical basis of the new method of bleaching various materials. G. G. Stokes demonstrated that many substances after light absorption emit intensive radiation, without any chemical changes taking place in their molecules. This phenomenon is called fluorescence or photoluminescence. In essence this results from the fact that the energy which is set free on the return of the excited molecule into its ground state is irradiated as a photon. The emission spectrum of such a substance appears as a broad band (approximately as broad as the absorption band, i.e. about 100 nm) which is shifted, however, to longer wavelengths. This general observation is known as Stokes' first rule. According to it the frequency of the light emitted by fluorescence must always be lower than the frequency of the absorbed light, i.e. a substance capable of fluorescence will emit fluorescence radiation of a longer wavelength than the light which is absorbed by the fluorescing substance. G. G. Stokes defined a further three rules for fluorescence, but this first one was of greatest importance for the gradual development of optical brightening agents.

In agreement with this rule, V. Lagorio discovered in 1921 that fluorescing dyes remit more visible light than corresponds to the absorbed visible light and he came to the conclusion that the reason for this phenomenon is the transformation of part of the light from the ultraviolet region into visible light. In 1929 Paul Krais carried out the first practical experiment by brightening a linen cloth with a natural optical brightening agent. Krais observed the strong fluorescence of common horse-chestnuts in ultraviolet light. Extracting them he obtained the substance causing this fluorescence, i.e. aesculin, a glucoside of 6,7-dihydroxycoumarin. A linen cloth dyed with aesculin displayed a high brightness, but in light the coloration changed rapidly to a yellow-brown hue. In his paper "Über ein neues Schwarz und ein neues Weiss"[1] he concluded with the historical sentence, that "the so far whitest white can be made still whiter...", which for many years has occupied the technical goal of numerous chemists and colorists.

In 1934 collaborators of the English firm Imperial Chemical Industries (ICI) prepared the first synthetic brightening agent, a diacyl derivative of 4,4'-diaminostilbene-2,2'-disulphonic acid, although it is not known whether it was practically tested or not. A year later, in 1935, Hoffmannsche Stärkefabrik Ultracell GmbH tested a further natural preparation – umbelliferoneacetic acid. Thus were the basic fluorescence systems first discovered, which are still important and which have been modified and substituted in the most various manners, achieving steadily increasing importance. In 1937 derivatives of umbelliferoneacetic acid began to be used as optical filters and protecting agents in the food industry. Before that the firm Kopp and Joseph introduced the production of coumarin derivatives as light filters for protecting papers for food products. In 1940 IG Farben introduced a small palette of optical brightening agents based on 4,4'-diaminostilbene-2,2'-disulphonic acid (Blancophor B, Blancophor R) and on sulphonated 4,5-diphenylimid-azolone (Blancophor WT). The success of these agents induced a rapid development of optical brighteners. Innumerable preparations were described, of which, however, the majority did not even attain the production stage and a number of those that did, had only a short commercial life, since the requirements put on the quality and use of optical brighteners constantly increased. Products of a higher stability were demanded, as were further products that would whiten materials other than cotton, such as silk, wool, paper, semisynthetic and synthetic materials.

In 1942 the Ciba company put on the market optical brighteners for cotton and polyamides, based on bis-benzimidazole (1).

(1)

In 1943 CIBA carried out research based on the first German palette of optical brightening agents and gradually introduced a broad

---

[1] On a new black and a new white

assortment of optical brighteners based on 4,4'-diaminostilbene-2,2'-di-sulphonic acid (the so-called CC/DAS preparations, formed by condensa-tion of two molecules of cyanuric chloride with one molecule of diamino-stilbenedisulphonic acid). Their relatively easy preparation by condensa-tion of cyanuric chloride with 4,4'-diaminostilbene-2,2'-disulphonic acid and a further aromatic or aliphatic component, and the excellent applica-tion properties of the products obtained led to the creation of many new optical brighteners (2).

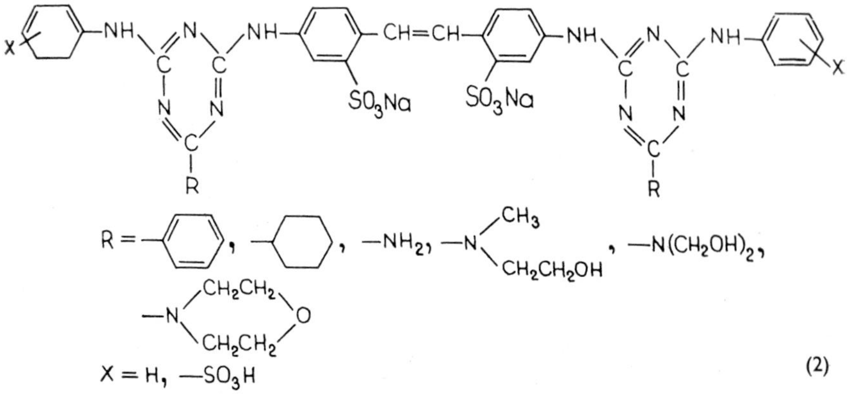

$$X = H, \ -SO_3H \tag{2}$$

The agents of this type are water-soluble, they possess a good affinity for fibres (especially cellulose fibres), excellent whitening properties, and they are stable in alkalies and during chemical bleaching. They are used mainly in the form of additives to detergent mixtures, for brightening cotton and linen goods. In 1944 IG Farben produced derivatives of 4,4'-bis(aroylamino)stilbene-2,2'-disulphonic acid used as brighteners for washing powders (3).

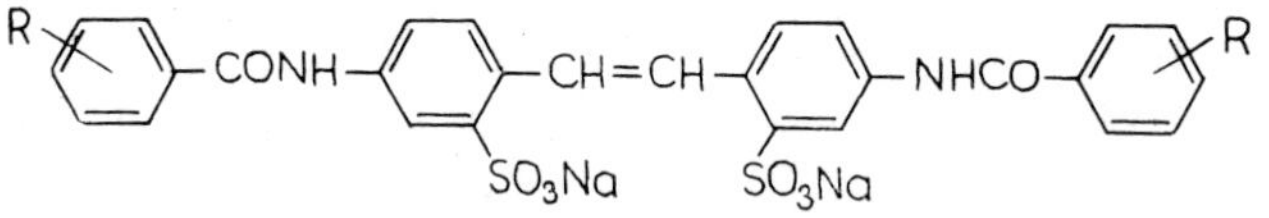

After the Second World War the development of optical brighteners greatly accelerated. The results achieved by IG Farben served as a starting point for the Swiss firms Ciba, Geigy and Sandoz, and then for the large British, USA and Japanese dye-producing firms. Other types of com-pounds, different from the stilbene ones, were looked for. In 1945 Ciba developed, for brightening plastics and synthetic fabric, derivatives of bis-benzoxazole (4), and, shortly afterwards, derivatives of 7-amino-coumarin (5) for whitening wool and polyamide fibres (1946).

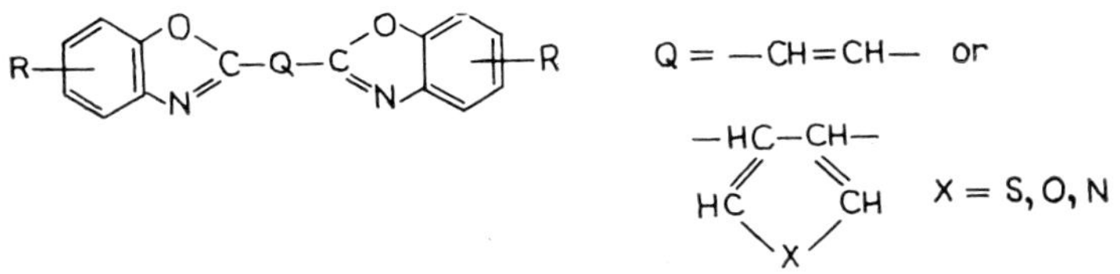

In 1948 optical brightening agents were developed based on 1,4-bis-(styryl)benzene (6), used for the brightening of cotton, wool and synthetic fibres, as well as 4,4'-bis-(naphthotriazolyl)stilbene (7).

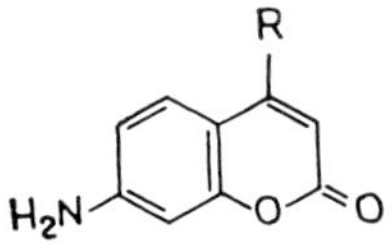

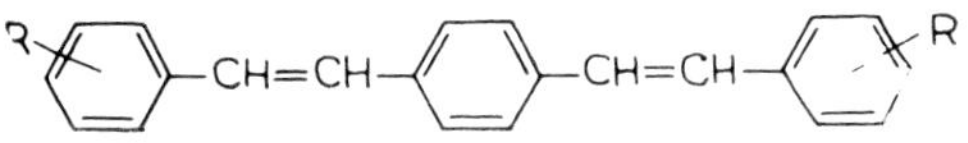

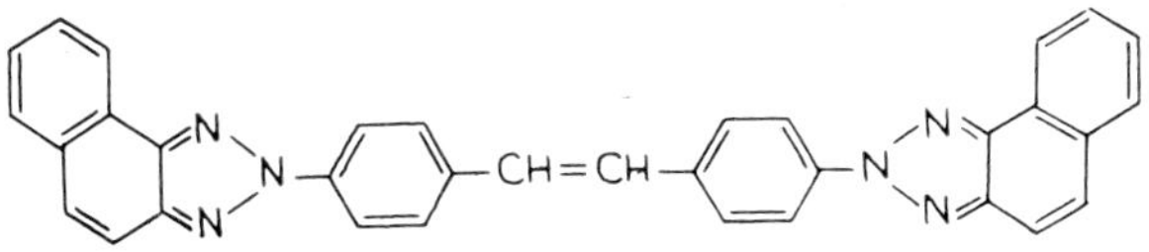

In the following year (1949) Ilford discovered a new group of optical brightening agents for polyamide and polyacrylonitrile, based on oxacyanins or more usefully on pyrazoline (8). The interest of dye producers very soon concentrated on this latter group of optical brighteners.

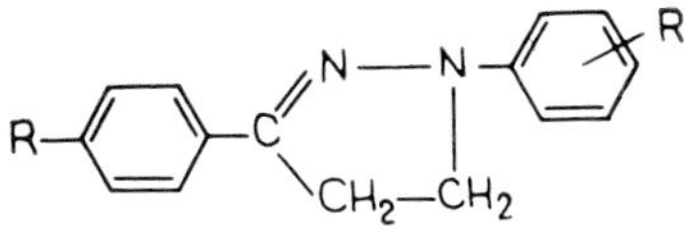

In 1951 Geigy put on the market stilbyltriazoles (9) and in 1952 BASF introduced naphthalic acid imides; in 1954 Geigy came up with derivatives of 3-phenyl-7-aminocoumarin (10) and later (1957) with derivatives of pyrazine (11).

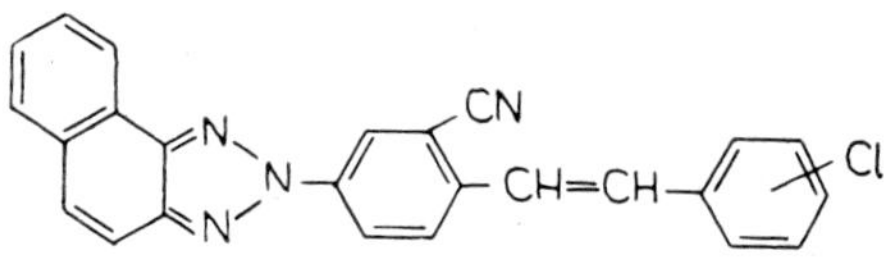

(9)

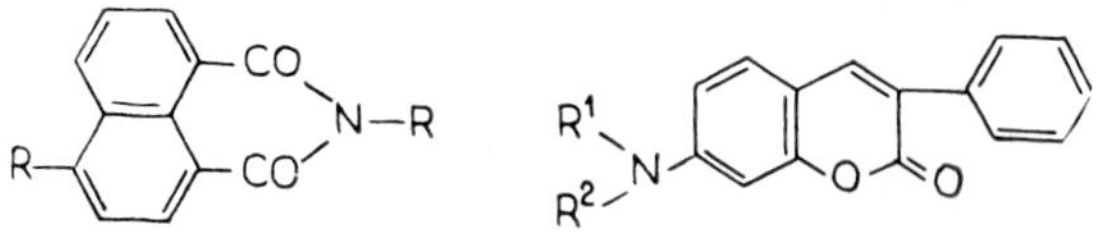

(10)

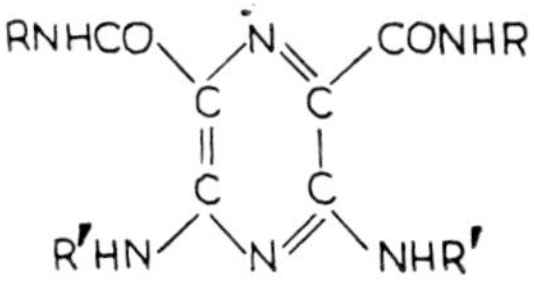

(11)

In the following years new derivatives of the groups already mentioned as well as other substances belonging to the most varied chemical areas were sought. In the still active area of optical brightening agents, interest is particularly centred on those based on 4,4′-diamino-stilbene-2,2′-disulphonic acid, coumarin, pyrazoline, styryloxazole, distyrylbenzene, naphthalenedicarboxylic acids and heterocyclic acids.

The demands placed on optical brighteners are steadily increasing. The range of application of these agents is becoming wider, the requirements for quality and purity are increasing, views on the shade of the agents are changing and a universally applicable substance is being sought.[1]

After 1955 other countries also began to investigate the development of optical brightening agents, for example the USSR, Poland, the German Democratic Republic, Czechoslovakia, and most recently Japan.

Optical brighteners today represent an integral part of the products of all the most important dye producers. The world market now carries more than 2500 trademarks, representing 200 various products belonging to more than 15 structural types. More than 30 companies produce optical brightening agents.

The use of optical brighteners is gradually penetrating a number of other industrial branches. From the textile industry, where they serve to brighten all natural and synthetic materials, they have passed into the

---

[1] The main interest today in the development of optical brightening agents is the search for preparations for the brightening of polyamides and polyesters or other types of synthetic fibres.

**Table 1 Consumption of optical brightening agents
in individual industrial branches**

Detergent mixtures	40 %
Paper	30 %
Synthetic fibres and plastics	5 %
Textiles	25 %

Principal world producers of optical brightening agents

Producer	Name of the product
Acna, Italy	Citalka
Allied Chemical, USA	Fluorescent
	Fluorosol
Amar Dye Co., Ltd., India	Amar White
American Cyanamide Company, USA	Branco Calcofluor
	Calcofluorwhite
	Fluorwhite
BASF[1], FRG	Blankit, Blankofil,
	Ultraphor, Palanilweiss,
	Palanilbrillantweiss,
	Ultraweiss
Bayer, FRG	Blankophor
Bulgaria	Belotex
Ciba, Switzerland	Uvitex
Daiton Kagaku, Japan	Daitophor
Du Pont, USA	Pontamine White
	Safariton White
	Paper White, Suowell
Durand and Huguenin, Switzerland	Albaphon
Dyestuffs Chemical Ltd., Great Britain	Lumisol
Eastman Kodak, USA	Brightener
Estab, Nac. Ind. de An., Brazil	White Enianil
Farb. Nac. y Expl., Spain	Blancophor
Francolor, France	Fluotex
Farbwerke Höchst, FRG	Hostalux
Farbenfabriken Delft, Netherlands	Delftwhite-, Delftweiss-
	Blancophor, Tintofen,
	Antara brightener,
	Brightener
Gujarat Chem. Ind., India	Flocin
Geigy, Switzerland	Tinopal
General Anilin, USA	Fluorol

Table 1 (continued)

Producer	Name of the product
General Anilin, USA	Fluorol
Hilton-Davis Chem. Co., USA	Arctic White,
	Hiltamine Mirawhite
Holliday Co. Ltd., Great Britain	Blancol
Hilton-Davis/Lamberti, Italy	Blancotex
Hickson and Welch, Great Britain	Photin
Hickson and Dadajee privat Ltd., India	Photin, Solium
Chimosa, Italy	Chimiflor
ICI[2], Great Britain	Fluolite
Mitsui Kayaku, Japan	Mikephor
Meotti, Ind. Chim., Italy	Miveal
Hungary	Optinol
Nippon Kayaku, Japan	Kayakoll, Mikowhite,
	Kayalite, Kayaphor
Nihon Soda K. K., Japan	Kayakoll
Nied. Farb. a. Chem. Fabrik, Netherlands	Delft Weiss
Origo, Italy	Bianco ottico
Pharma-Chem. Corp., USA	Phorwhite
Poland	Heliophor, Hekkol
Roumania	Stralex, Albaton
Sandoz, Switzerland	Leukophor, Leucophor
	Lucopur, Sandowhite
Showa Kayaku, Japan	Hakkol
S. A. Rovira, Bachs and Macia, Spain	Photin
Societé des Produits Chimiques, France	Celumyl
Sumitomo, Japan	Whitex
Sigma, Italy	Optiblanc
USSR	Belyj, Belophor
Unichem, Czechoslovakia	Rylux
VEB Chemiekombinat Bitterfeld, GDR	Weisstöner Wolfen,
	Wobital-
Verona Pharma Chemical Co., USA	Phorweiss

[1] Badische Anilin- und Sodafabrik
[2] Imperial Chemical Industries

paper industry, where they are used to whiten paper; in the photographic industry they are added to developers. They have also found application in the leather-working industry and in the production of plastics. Thus, like dyes, optical brightening agents have become a part of everyday life and culture.

1.2 Statistical data on world production

The production and the consumption of optical brighteners is constantly increasing and in recent years the annual increase has amounted to 10 – 12 %. The world production of these agents at present is estimated at more than 120 thousand tons per year. In 1972 in the USA the share of optical brightening agents was 10.4 % of the total production of dyes, while in 1974 it was 14.9 %.

As consumers of optical brightening agents, the laundry industry takes first place, the paper industry second place, while the third and the fourth places are shared by the textile industry and the fibres and plastics industries. The distribution of consumption of optical brightening agents in individual industrial production branches is shown in Table 1.

1.3 Theoretical basis of fluorescence

1.3.1 Fundamental relationships in the absorption of light and colour

The colour of an object, perceived by the human eye, is determined by that part of the total spectrum of white light transmitted or absorbed by the object. Sunlight is composed of ultraviolet, visible and infrared radiation, ultraviolet and infrared light being invisible to the human eye. The visible part of daylight can be resolved into a spectrum of several colours: violet, blue, green, yellow, orange and red. The violet end of the visible spectrum is extended to the ultraviolet region, the red end is extended to the infrared region. Individual regions or colours of resolved daylight correspond to different wavelengths of the light rays:

Wavelength (nm)	Region of light	Complementary colour
up to 400	ultraviolet	
400 to 424	visible-violet	yellow-green
424 to 492	blue	yellow
492 to 565	green	red
565 to 585	yellow	violet
585 to 647	orange	blue
647 to 760	red	green
above 760	infrared	

The incident white light can be completely absorbed by the object or completely reflected. In the first case the object appears black (total absorption), in the second case it appears white (total reflexion). If the object absorbs only a part of the incident white light, our eye catches only the remaining part of the spectrum, reflected or transmitted by the object, which appears to us coloured. We perceive the reflected or the transmitted part of light as the colour complementary to the absorbed part of the light. Hence, the colour of the object perceived by our eye is not spectrally uniform or pure, but mixed. Therefore the same colour can result from a different composition of the transmitted or the reflected light. The light absorbed or reflected by the object is characterized unambiguously by its absorption or reflectance spectral curve.

The absorption zones of the various colours in the spectrum of decomposed daylight are not sharply defined. An absorption maximum — dependent on the structure of the dye — passes on both sides into a characteristic spectral curve that represents the distribution of the light absorption of the dye at various wavelengths. Dyes of brown and red hues do not have maxima with sharply dropping absorption curves, but they absorb in a broad region and are characterized by an extensive absorption band stretching over a broad range of wavelengths. If the perceived colour of an object is shifted from yellow to violet, i.e. its absorption in the direction of longer wavelengths, we speak of a *deepening* of the colour, of a *bathochromic* shift, or of a (so-called) "red shift", while if the colour changes from violet to yellow, i.e. in the direction of shorter wavelengths, we speak of a *lightening* of the colour, of a *hypsochromic* shift, or of a (so-called) "blue shift".

In addition to the hue, each dye also has a certain intensity. The increase of the intensity produced by various structural factors is called the *hyperchromic* effect and the decrease of the intensity is called the *hypochromic* effect. The position of the absorption maximum is characterized by its wavelength, frequency and wave-number. The wavelength λ determines the length of the wave of monochromatic light. Its basic unit is the metre; very often it is expressed in nanometres (nm), or less frequently centimetres, millimetres and Ångströms (1 nm $= 10^{-9}$ m $=$ $= 10$ Å). The relation of frequency and wavelength is expressed as $\nu\lambda = c$, where c is the velocity of light ($3 \cdot 10^8$ m s^{-1}), (or — when expressed in nanometres $\nu\lambda = 3 \cdot 10^{17}$ nm s^{-1}).

By wave-number $\tilde{\nu}$ we mean the number of waves per 1 m. The relationship between the wave-number and the wavelength is given by the expression $\tilde{v}\lambda = 1$. The unit for the wave-number is a reciprocal metre (m^{-1}), also called a Kayser. However, usually the wave-number is given in reciprocal centimetres (cm^{-1}).

A molecule does not absorb light energy continuously, but it always selects from the luminous flux a definite quantum of energy, a so-called photon. It is usually said that the luminous flux forms a discrete series of energy parts – photons, certain of which are selected from the luminous flux by the molecule. The energy of the photons is proportional to the wave-number and it is equal to hv, where h is Planck's constant $(h = 6.62 . 10^{-34} \, J \, s^{-1})$.

Light absorption follows the Lambert – Beer rule

$$A = \varepsilon_\lambda bC = \log \frac{\Phi_d}{\Phi_t}$$

where A is absorbance, C — molar concentration of the substance, b — thickness of the layer through which the ray passes, Φ_t — entering luminous flux (intensity of incident light), Φ_d — out-going luminous flux (intensity of transmitted light), ε_λ — molar absorption coefficient (formerly: molar extinction coefficient).

1.3.2 Absorption and emission of light by the dye

Dye molecules in the ground state of energy E_0 may be designated as A, while molecules brought by absorption of photons hv to a higher energy level E^* are called excited molecules and are designated by the symbol A*. In contrast to the ground state the excited state of the molecule is not stable and has a very short life time τ. After about 10^{-9} s the majority of excited molecules return to the ground state A and the molecule can then absorb another light quantum. Many dyes when dissolved in liquid or solid substances absorb light and emit radiation strongly without any chemical changes taking place. This effect is called fluorescence. It is caused by the irradiation of the energy set free on the return of the excited molecules to their ground state. Both processes, absorption and emisson of light energy hv (photon) can be summarised by the equations:

absorption:

$$hv_a + A \;\;\rightarrow\;\; A^*$$

emission:

$$A^* \;\;\rightarrow\;\; A + hv_{fl}$$

The wavelengths which are absorbed and irradiated by the dye cover a relatively broad region of about 100 nm, but the emitted light is always shifted towards longer wavelengths than those of the absorbed light. This feature of fluorescence is known as Stokes' first law, according to which the frequency of the emitted light must always have a lower value then the frequency of the absorbed light:

$$hv_{fl} < hv_a$$

The residue of the energy, given by the difference in the values of the absorbed and released light energy is retained in the molecule as thermal energy, transmitted to the environment as heat.

A characteristic feature of the spectral fluorescence band is its asymmetric shape, with a very sharp drop on the long wavelength side, i.e. the opposite of the absorption curve. Spectral curves for absorption and fluorescence are like mirror images. The rule of mirror symmetry of the absorption and the fluorescence bands is called Stokes' second rule.

Stokes' third rule states that when the substance is excited by monochromatic light in any part of the spectrum, the whole fluorescence spectrum is emitted, and is unchanged in its wavelength distribution and intensity. (When excitation is caused by light of wavelength not corresponding to the absorption band maximum, the total intensity of the fluorescence band decreases correspondingly.) The cause of this phenomenon evidently arises from the fact that the absorption of photons of different energy hv_a brings the molecule from the ground state to several different energy levels of the excited state. On monochromatic excitation only a single level is attained. If simultaneously a whole set of photons hv_{fl} is emitted with an unchanged intensity distribution, it means that between the act of absorption and the act of radiation deactivation of the excited molecule takes place, leading to a single energy level of the excited state, most suitable for emission. This normalization of the excited states is an intra-molecular property of the molecule and it is not caused by the interaction of the molecule with its surroundings.

The fourth rule of fluorescence is the universal mirror image relationship between the fluorescence band and the absorption band of a substance. The relationship arises from the similar spacing of vibrational levels in the excited state (controlling the shape of the absorption band) and the ground state (controlling the shape of the fluorescence band).

The ability of a molecule to emit the absorbed energy as light energy is measured by the quantum yield of fluorescence φ_{fl}, which is the ratio of the number of photons $h\nu_{fl}$ (or molecules), emitted within the whole range of the emission band, and the number of photons (or molecules) absorbed within the same time interval on monochromatic excitation with light of a certain frequency ν_a.

The quantum yield of fluorescence depends on the structure of the compound and also on the medium and temperature.

Dye molecules that are dissolved in a solid medium may display, in addition to a weak fluorescence, a longer lasting additional radiation, called *phosphorescence*, the emitted band of which is shifted to longer wavelengths. This effect is most pronounced at low temperatures. Hence, the absorbed energy can be emitted intramolecularly in two ways: by fluorescence and phosphorescence.[1] Therefore, by analogy with fluorescence, the quantum yield of phosphorescence may also be defined as the ratio of the number of photons $h\nu_{ph}$ emitted within the whole range of the spectral band of phosphorescence to the number of photons absorbed within the same time on monochromatic excitation with frequency ν_a.

1.3.3 Theory of fluorescence

The energy of a molecule is given by the sum of the electronic energy, vibrational energy, rotational and kinetic energy (which is negligible owing to its low value).

The vibrational movements of the atoms are very complex. Vibrational energy increases slightly under the effect of intermolecular collisions and mainly with increasing temperature. The increase or decrease of vibrational energy follows quantum laws, i.e. the transition of one vibrational level to another. Under the conditions of thermal equilibrium at temperature T energy RT can be assigned to the set of independent types of vibrations in the molecule, where R is the molar gas constant of value $8.3 \text{ J mol}^{-1} \text{ K}^{-1}$. At room temperature (about 290 K) this means a value of about $250 \text{ J mol}^{-1} \text{ K}^{-1}$. In its ground electronic state the molecule already possesses a certain amount of vibrational energy E_V, created by the thermal process. At temperature T the molecules are on average at the energy level E_V. On electronic excitation from the level E_e^0 to the level E_e^* a change in vibrational energy takes place as a rule, i.e. it increases by

---

[1] Both phenomena are collectively called luminescence.

ΔE_V. Hence, the absorption of a light quantum is higher by this value, i.e. $h\nu_a = h\nu_0 + \Delta E_V$.

In Fig. 1 individual lines represent the levels of vibrational energy E_V both at the lowest electron level E_e^0 and the highest electron level E_e^*.

In dye molecules the intervals between individual levels E_V are equal. In large dye molecules the unoccupied levels are filled on light absorption up to the total vibrational energy E_V, which appears in the spectrum as a set of closely agglomerated lines, i.e. an absorption band

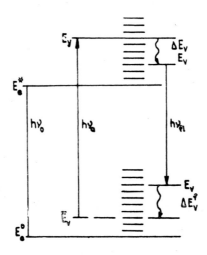

Fig. 1. Transitions taking place in the molecule between electronic (E_e) and vibrational (E_v) energy levels on absorption and emission of light

is formed. On deep cooling of dye solutions certain vibrational energy levels may disappear, thus causing the splitting of absorption or luminescence bands to a fine structure, reproducing the order of vibrational levels (Spolsky's effect). The gradual decrease of absorption in the band in the direction of higher frequencies indicates that an increase in vibrational energy on the absorption of a photon cannot take place without limitation.

Similarly to light absorption, light emission also takes place. A part of the vibrational energy ΔE_V (i.e. the same amount by which the vibrational energy increased during absorption) is distributed between the

numerous vibrations of the molecule, or it is transmitted to the environment, so that an equilibrated distribution of energy takes place, corresponding to the upper state of the level E_V. If the substance is fluorescing
the majority of the molecules emit light from this level. Similarly as
during absorption, vibrational energy also increases during light emission,
by the value ΔE_V^f, so that the emitted quantum $h\nu_{fl}$ is less than the absorbed
quantum $h\nu_a$ by the value of the total residual energy

$$h\nu_a - h\nu_f = E_V + \Delta E_V^f$$

The first part of the increase of vibrational energy, ΔE_V, is lost
by the molecule before the beginning of the emission, while the second
part, ΔE_V^f, is formed in the molecule at the moment of the emission.
Vibrational energy increases with absorption and its excess is converted
into thermal energy. This process is called degradation of the excitation
energy to heat.

Some substances emit fluorescence radiation which has a higher
energy than that absorbed (anti-Stokes substances).

1.3.4   Phosphorescence and delayed fluorescence

Simultaneously with fluorescence which lasts 10^{-7} to 10^{-9} seconds,
a further type of luminiscent radiation with a longer radiation time than
fluorescence appears after switching off the excitation source — called
phosphorescence. Phosphorescence lasts up to several seconds after
removing the excitation radiation, and it is prolonged when the molecule
of the emitting substance is in a solid organic or inorganic medium (as for
example boric acid, sugar ice, cellulose, gelatine), especially when such
substances are cooled to low temperatures (77 K). At room temperature
the luminiscence band of the substance in solid solution may be identical
to the band of normal fluorescence, and it differs from it merely by the
anomalously long life-time, 10^5 to 10^9 times longer than normal
fluorescence. Such persistent radiation is called *delayed fluorescence*.
In addition to the delayed fluorescence band, a new band may appear
in the emitted spectrum which rapidly increases at low temperatures,
and which eventually replaces the former band. The new band, which is

phosphorescence, is shifted in comparison with the former to longer wavelengths. Both emission bands, delayed fluorescence and phosphorescence, are produced by excitation of the long wavelength band of the dye. At low temperatures their life-time is several seconds and their emission bands show fine structure.

1.4 Relationship between structure and fluorescence

Several theories have been elaborated concerning the relationship between the structure and fluorescence ability of substances.

The oldest, so-called classical theory, does not substantially differ from the auxochromic theory of colour, introduced in 1876 by O. Witt. According to the latter theory the carriers of colour are the double bonds, the so-called *chromophores*. If introduced into colourless substances, they may render them coloured. Such chromophores are, for example, the azo, nitro, nitroso, and carboxyl groups. A substance containing chromophores is called a chromogen. Hence, a chromogen is an already coloured compound containing one or more unsaturated chromophoric groups. However, such a substance cannot in most cases be considered a dye (with a few exceptions). Chromogens usually lack the necessary colour intensity and affinity for fibres. Both properties are obtained after introduction of further groups into the molecule, for example hydroxy, amino, alkylamino and acetylamino and similar groups. These groups are called *auxochromes*. Accumulation of chromophores (such as extension of the chain of conjugated double bonds, fusion with aromatic rings, or accumulation of mutually conjugated oxo groups, etc.) causes a bathochromic shift, i.e. absorption is shifted in the direction of longer wavelengths. Since the colour of a substance is caused by a number of chromophores it is more correct to describe such compounds as *chromophore systems* rather than as chromophores only. Auxochromes affect not only the intensity of the colour, but also the depth of the colour.

Dissociation of hydroxy groups in such a substance (formation of phenoxide salts) has a bathochromic effect, whilst protonation of amino groups causes a hypsochromic effect. Alkylation and arylation of amino groups causes a bathochromic shift, while similar substitution of hydroxyl groups causes a small or zero hypsochromic effect. Acylation of amino groups produces a hypsochromic shift. Chromophores and auxochromes must exist in a specific mutual relationship for maximum colour production.

A certain auxochrome, if combined with one chromophore, can intensify the colouration, while with another this can be weakened. Even the same auxochrome and the same chromophore can behave differently on changing their mutual positions in the chromogen. Therefore a chromophoric and an auxochromic system should always be considered in mutual relationship, so that often the literature speaks about an auxochromophoric system, meaning the combination of the chromophoric and the auxochromic systems. Although Witt's theory cannot explain all the phenomena of colouration, it is still very much used, owing to its simplicity and clarity. For this reason it is also applied in the area of fluorescing substances.

Aromatic hydrocarbons such as benzene, naphthalene, anthracene and fluorene are called luminophores and are considered as basic fluorescing substances. The groups that enhance this fluorescence, or that shift it to the visible region, for example $-CH=CH$, $-CO-$, p-phenylene, $-CH=CH-COOH$ or $-CN$, are called fluorogens. A luminophore carrying a fluorogen is called a fluorophore. Optical and coloristic properties of brightening agents are affected — similarly as in dyes — by auxochromes or antiauxochromes, which are sometimes called, in the case of fluorescing substances, auxoflores and diminoflores. Auxoflores include $-NH_2$ and $-OH$ groups which enhance fluorescence. Diminoflores include for example the $-SO_3H$ and $-COOH$ groups which decrease fluorescence. The effects of auxoflores and diminoflores are not unambiguous and depend on the position they assume in the molecule. If an auxochrome (for example $-NH_2$) and an antiauxochrome (for example $-NO_2$) are mutually in the p-positions of a benzene ring, the dipole moment is increased and the intensity of fluorescence decreased; in contrast to this, if two auxochromes are mutually in the p-position, the intensity of fluorescence increases because the dipole moment of the molecule is small. Auxoflores (auxochromes) will enhance or weaken fluorescence depending on whether their mutual position (or in combination with diminoflores) will cause a decrease or an increase in the dipole moment of the molecule. The larger the dipole moment of the molecule, the more pronounced will be the weakening of the fluorescence.

Mumm's theory helps to answer the question of the relationship between structure and fluorescence by means of excitation-induced mesomerism between three neighbouring atoms in the general grouping

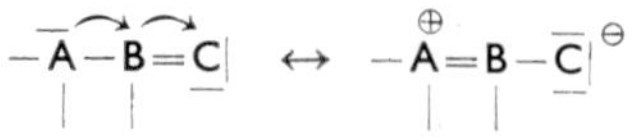

$$-\overline{A}-B\!=\!C| \quad \leftrightarrow \quad -\overset{\oplus}{A}\!=\!B-\overline{C}|^{\ominus}$$

The three atoms A, B and C are bound by one pair of electrons, i.e. by two π-electrons. If such a system absorbs light, this can be achieved solely by these π-electrons. When this three-atom system is excited a mesomeric shift takes place to one of the limiting resonance structures. Once this strained system is formed, the excited electron reverts back to the atom at the other end of the system, and there is emission of light energy. Mumm provided evidence for this theory experimentally with N-methyl-α-pyridone (12-I) which fluoresces if substituted in positions 3 and 5 with any substituent, and he assumed that fluorescence must be due to the $CH_3-N-CO-$ grouping. As soon as the pyridone structure has changed, fluorescence also disappears, as for example in the case of 1,4,6-trimethyl-5-carboxyethyl-2,3-cyclobutanone-2-hydroxydihydropyridine (12-II).

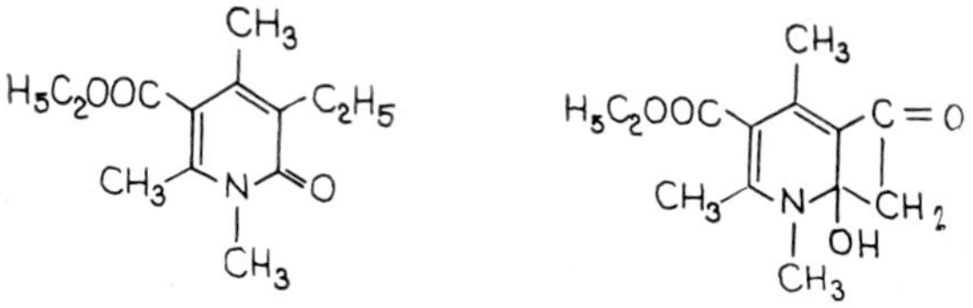

(12)

I   II

A similar situation is also found in γ-dihydropyridine and α-dihydropyridine, where the γ-derivatives (13-I) fluoresce strongly, while the α-derivatives (13-II) do not fluoresce.

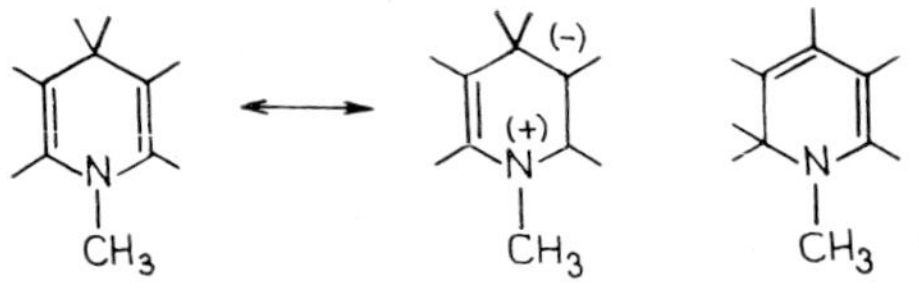

(13)

I   II

Didier's oscillation theory assumes that the originator of fluorescence is the atom of nitrogen or oxygen, while the rest of the molecule functions as an electromagnetic oscillator which has its own absorption with corresponding fluorescence.

Zelinskij proposed a similar theory as Mumm, assuming that localization or delocalization of the electron charge in the molecule takes place under the effect of the chromophore system, so that fluorescence does not take place when a strong dipole moment is formed.

1.5 Classification and nomenclature of optical brightening agents

As in the case of dyes, the classification of optical brighteners can be based either on the chemical structure of the brightener or on its method of application.

From the point of view of their use, optical brighteners are classified primarily into two large groups: direct (or substantive) brighteners and disperse brighteners.

Direct optical brightening agents[1] are predominantly water-soluble substances, used mainly for the brightening of natural fibres and occasionally for synthetic materials such as polyamides.

Disperse optical brightening agents are mainly water insoluble, and as with disperse dyes they are applied either to coloured material from an aqueous dispersion, or they can be used for mass colouration. They are used mainly for synthetic materials, for example polyamide, polyester, polyacrylonitrile, acetate silk, polyvinyl chloride and occasionally natural substrates (e.g. paper).

In addition to this basic classification of optical brighteners, individual types are also labelled according to the material for which they are intended (optical brightening agents for cotton, wool, polyamide, polyester, etc.).

From the chemical point of view optical brighteners are classified according to their chemical structure. This classification is not rigid and depends on how detailed a view is taken of the very wide range of chemically different types of optical brightening agents. From the chemical viewpoint optical brightening agents are classified into derivatives of stilbene, coumarin, 1,3-diphenylpyrazoline, derivatives of naphthalenedicarboxylic acids, derivatives of heterocyclic dicarboxylic acids, derivatives of cinnamic acid, and substances belonging to other chemical systems.

The commercial names of individual brands are variable and unsystematic. Usually the name of the optical brightening agent does not give a clue to the application similarity of the brightening agents to a related class of dyes. The letters after the name of a brightening agent do not uniformly express the characteristics of individual brands.

---

[1] The adjective "direct" is not usually quoted. The general term optical brightening agent refers almost always to direct preparations.

In the Soviet Union a system has been introduced where the first letter after the name of the brightening agent indicates the shade (K — reddish, S — bluish, O — neutral, Z — greenish) and the following letters indicate to which material the agent can be applied (B — paper, D — washing agent, L — polyester, N — polyacrylonitrile, P — polyamide, M — brightening in the mass, S — wool, silk, V — viscose, A — acetate, triacetate). The multiplicity of the shade is expressed by a number.

1.6   Mechanism of whitening with optical brighteners

The large majority of organic materials, either natural or artificial textile fibres or other substances, such as paper, plastic, etc., do not look completely white. Generally they absorb slightly more violet radiation (400 to 480 nm), than yellow radiation, and in consequence they appear more or less yellowish. The yellowish hue of white materials can be eliminated both by chemical bleaching and by blueing or whitening with optical brightening agents.

Chemical bleaching with any type of agent eliminates some of the light absorbing impurities and components of the material, but perfect whitening cannot be achieved with this method.

Blueing, carried out by dyeing the materials with a blue inorganic pigment or even with blue organic dyes produces both a compensation of the yellow colour of the fibre and a blue coloration (subtractive process), achieving thus a whiter impression for the dye. In addition to the physical process of the compensation of two dyes, it is the physiological impression which plays an important role here. Our eye is used to perceiving blue or greenish hues as white[1]. Hence, blueing brings about a whiter shade, but since the compensating blue dye itself also absorbs in visible light, the total amount of reflected light is smaller than in the case of the unblued material, so that the whitened material is less bright, i.e. it is dull, or greyish.

The most efficient whitening is achieved by dyeing the respective materials with optical whitening agents.

The term "dyeing" suggests that optical whitening agents are regarded as dyes. Optical whitening agents also have their own characteristic shade (as dyes have) — blue, violet, green, red — although of very

---

[1] In Europe blue and violet shades of white are preferred, whereas in America a greenish hue is preferred.

low intensity. As in the case of dyes, they absorb a part of incident daylight, i.e. the invisible, ultraviolet component, and emit this light energy — as we have seen in the preceding text — as long wave visible light in the blue part of the visible region (an additive process). Hence, they compensate by their emission colour for the yellowish shade of the material, but at the same time the brightness of the dyed material is increased through the fact that the eye perceives in fact a larger amount of visible light reflected by the material than was incident on it. The material appears whiter and brighter. The theory and mechanism of luminescence processes have been explained in the preceding text.

An ideal optical brightener absorbs in that part of the ultraviolet spectrum which is closest to the visible spectrum and emits the absorbed light energy in the short-wave region of the visible spectrum as fluorescent radiation of various hues. According to Stokes' laws the shape of the fluorescence band can be predicted from the shape and the position of the absorption band, and the colour of the fluorescence can thus be determined. A violet fluorescence corresponds to a fluorescence maximum at $\lambda = 415$ to 429 nm, for $\lambda = 430$ to 440 nm it is blue and for $\lambda = 441$ to 466 nm it is greenish-blue. In addition to the localization of the maximum, the relative distribution of the emitted light within the emission band is also important. The peak of fluorescence lies within very narrow limits of wavelengths (about 10 nm). An ideal optical brightening agent should possess a strong absorption at $\lambda = 350$ nm, maximally shifted to $\lambda = 400$ nm; the absorption band should drop steeply into the visible region and the absorbed energy must be transformed to fluorescence as quantitatively as possible, so that the quantum yield of fluorescence should be close to unity. When brightening agents are applied to substrates the quantum yields are not equal to unity because of other competing intermolecular processes, causing a partial or total "quenching" of fluorescence. Similar quenching can also be produced by other substances, occurring as impurities either in the optical brightening agent itself, or in the dyed substrate.

2

Production of individual types
of optical brighteners

2.1 Stilbene compounds

About 80 % of all optical brightening agents produced are derived from stilbene (14), the latter absorbing in the ultraviolet region at $\lambda = 342$ nm. Introduction of suitable substituents into the 4 and 4′ positions shifts this absorption to longer wavelengths.

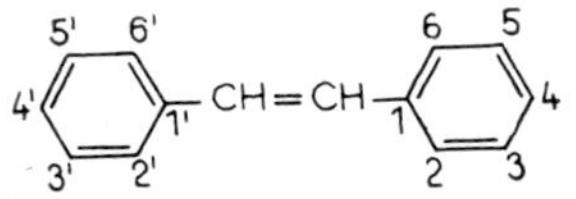

(14)

2.1.1 Derivatives of bis(triazinyl)aminostilbene

Derivatives of bis(triazinyl)aminostilbene are substances with excellent brightening properties and at the same time are easily produced. As basic starting substances for the production of optical brightening agents of this group 4,4′-diaminostilbene-2,2′-disulphonic acid[1] and 2,4,6-trichloro-1,3,5-triazine (cyanuric chloride) are used. Condensation of these two reagents and further combination of the condensation product formed with various aliphatic and aromatic hydroxy compounds or amino compounds gives a wealth of optical brightening agents, among which are many valuable, commercially exploitable compounds. They have in common high affinity and substantivity, and good colour fastness. 4,4′-Diaminostilbene-2,2′-disulphonic acid or its sodium salt (15-I) con-

---

[1] Optical brigteners prepared from 4,4′diaminostilbene-3,3′-disulphonic acid have more advantageous properties than those made from 2,2′-disulphonic derivative. However, the preparation of 4,4′-diaminostilbene-3,3′-disulphonic acid is still industrially uneconomic.

denses in an aqueous or aqueous-acetone medium and in the presence of substances that bind the released hydrogen chloride (for example sodium carbonate) with cyanuric chloride (15-II) to give 4,4'-bis(4,6-dichloro-1,3,5-triazin-2-yl)aminostilbene-2,2'-disulphonic acid (15-III). This intermediate is further condensed in two steps with amines, alcohols, phenols or thiols to give compounds of types 15-IV and 15-V.

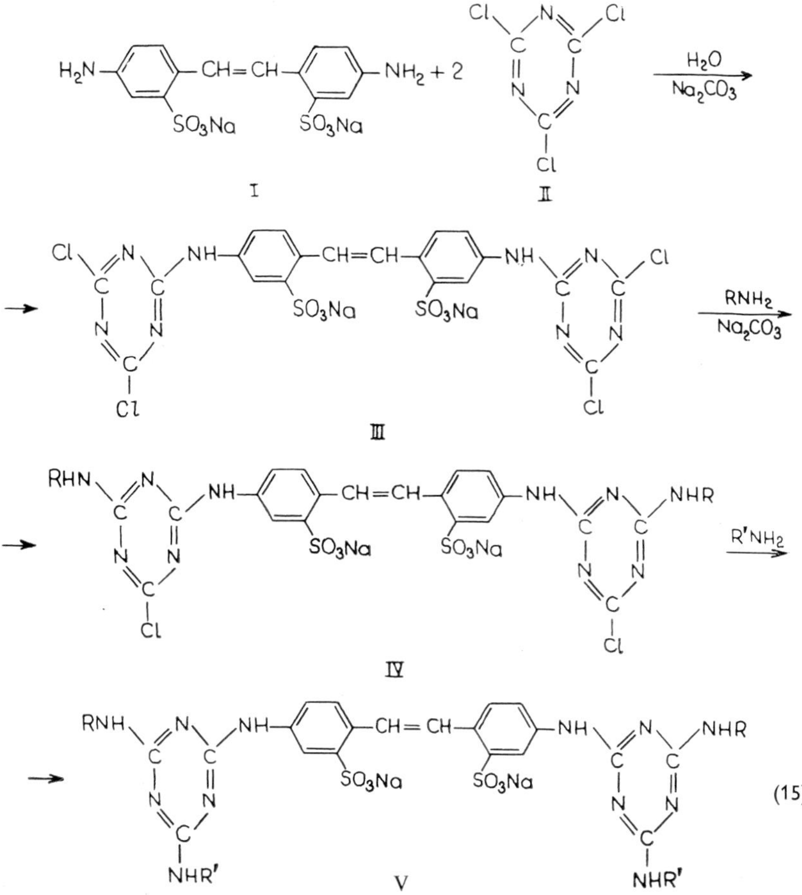

In this connection the reactivity of individual compounds should be taken into consideration. Cyanuric chloride is a trifunctional compound the three reactive chlorine atoms of which can only be replaced in a stepwise manner, the reactivity of each chlorine replaced decreasing at each replacement stage. At about 0 °C cyanuric chloride reacts as a monofunctional acid chloride that condenses, in the majority of cases, with the difunctional 4,4′-diaminostilbene-2,2′-disulphonic acid in a 2 : 1 molar ratio. It is obvious that this condensation is not exclusive and that other condensation products are also formed in the reaction, the absorption or the fluorescence of which can interfere with the emission curves of the required products. Condensation of 2-monosubstituted 4,6-dichloro-1,3,5-triazine with amines, alcohols or phenols must be carried out at higher temperatures (40 to 60 °C). The reaction is dependent almost exclusively on the first substituent of the triazine cycle and the reactivity decreases for the first substituent in the following order:

$$C_6H_5O > CH_3O > C_2H_5O > C_6H_5NH > CH_3NH >$$
$$> C_2H_5NH > (CH_3)_2N > OH.$$

The condensation with a third nucleophile takes place at 100 °C only and the rate dependence on substituents is similar to that described for secondary substitution. In dependence on the reactivity of all three components involved in the reaction the optical brightening substances are produced in three ways:

1. Cyanuric chloride is first condensed with methanol to give 2-methoxy-4,6-dichloro-1,3,5-triazine and the latter is condensed, without isolation, with diaminostilbenedisulphonic acid and an amine.

2. Cyanuric chloride is first condensed with diaminostilbenedisulphonic acid, and the intermediate obtained is condensed sequentially without isolation with further amino components.

3. Cyanuric chloride and the amine react to form a 2-arylamino-4,6-dichloro-1,3,5-triazine, which is isolated and condensed with diaminostilbenedisulphonic acid to give sodium 4,4′-bis(6-arylamino-4-chloro-1,3,5-triazin-2-yl) aminostilbene-2,2′-disulphonate. The latter is isolated and reacted with a further amine to give the final product.

Cyanuric chloride is prepared by polymerization of cyanogen chloride (15a).

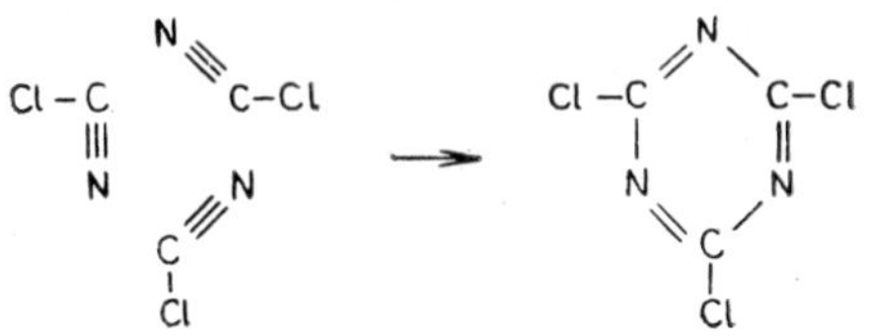

(15a)

From an aqueous solution of cyanogen chloride, obtained by chlorination of hydrogen cyanide, cyanogen chloride is distilled off, which is then dried and polymerized in benzene saturated with dry hydrogen chloride at 35 to 40 °C.

4,4′-Diaminostilbene-2,2′-disulphonic acid

This is prepared by three-step synthesis from *p*-nitrotoluene. The latter is first sulphonated with 26 % oleum at 55 – 60 °C to give 4-nitro-toluene-2-sulphonic acid (15b):

(15b)

The product is isolated, after transferring the sulphonation mixture into a larger amount of cooled water, by filtration.

A paste of 4-nitrotoluene-2-sulphonic acid is dissolved in a larger amount of water and neutralized with sodium hydroxide solution. The sodium salt of 4-nitrotoluene-2-sulphonic acid is then oxidized with a solution of sodium hypochlorite at 75 – 80° under simultaneous addition of sodium hydroxide solution.

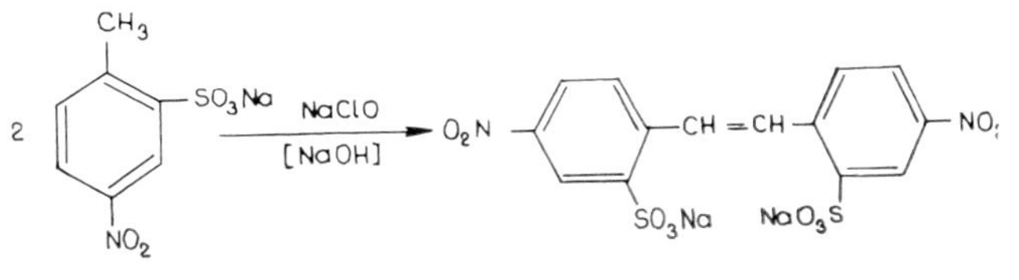

(15c)

After termination of the oxidation process the reaction mixture is neutralized with concentrated hydrochloric acid, cooled and salted out. The separated product is isolated by filtration on a filter press. The paste of 4,4′-dinitrostilbene-2,2′-disulphonic acid is then reduced by adding it gradually into an aqueous suspension of iron filings, etched with hydrochloric acid, at a temperature of about 100 °C.

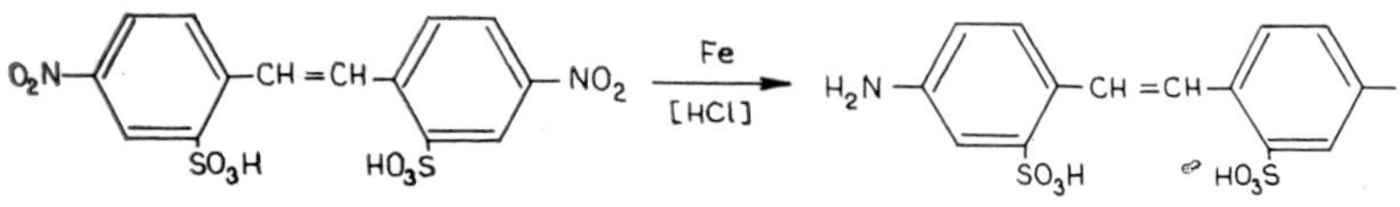

$$(15d)$$

The reduction mixture is made alkaline with sodium hydroxide and freed of the iron-containing sludge on a filter press. From the filtrate 4,4'-di-aminostilbene-2,2'-disulphonic acid (15 d) is precipitated by acidification with sulphuric acid. The product is filtered off. For further work-up the product is usually used directly in the form of a paste.

For illustrating these principles, the preparations of some optical brightening agents based on 4,4'-diaminostilbene-2,2'-disulphonic acid are given in detail.

4,4'-Bis-(6-phenylamino-4-methoxy-1,3,5-triazin-2-yl)aminostilbene-2,2'-disulphonic acid ((16) — a brightener of the type Uvitex PRS)

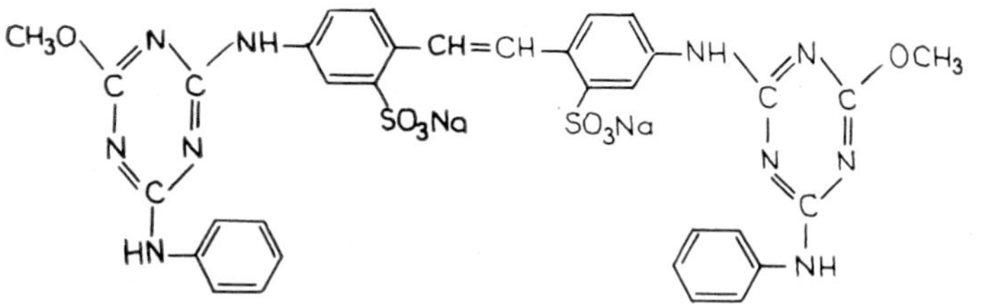

$$(16)$$

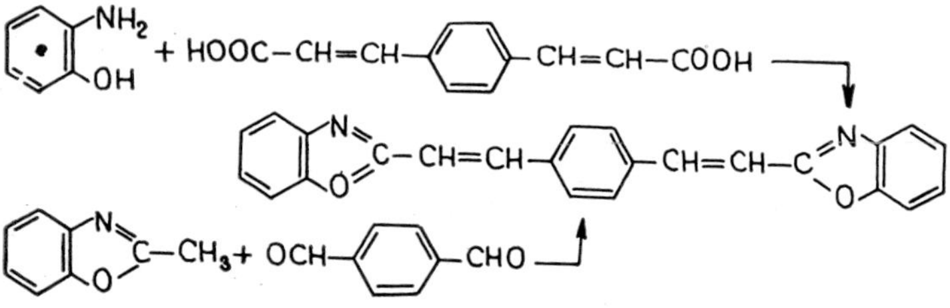

This substance is prepared by condensation of cyanuric chloride with methanol, 4,4'-diaminostilbene-2,2'-disulphonic acid and aniline. The production of the whitening agent involves the preparation of the sodium salt of the diaminostilbenedisulphonic acid and the actual condensation.

Preparation of the sodium salt
of diaminostilbenedisulphonic acid

Into a kettle filled with water diaminostilbenedisulphonic acid is introduced in the form of a paste with constant stirring. After adding alkali the suspension is stirred until the diaminostilbenedisulphonic acid is completely dissolved and the solution is weakly alkaline to Brilliant Yellow indicator.

Condensation

Into the condensation kettle methanol and water are introduced (from a container) and cooled with brine to 10 °C. Then cyanuric chloride is added into the kettle at once under thorough stirring. When thoroughly mixed the reaction mixture is cooled, alkali is added and the solution of the sodium salt of diaminostilbenesulphonic acid prepared earlier is added slowly. Towards the end of the addition the cooling is interrupted and the temperature is allowed to rise to 20 °C. The reaction mixture is diluted with water and heated to 35 – 40 °C. The condensation is terminated at this temperature within one to three hours, while keeping the reaction at pH 7. Aniline is then added to the suspension and after short stirring of the reaction mixture sodium hydrogen carbonate is added. The reaction mixture is then heated to 80 to 90 °C with constant stirring and the contents of the kettle are pressed through a filter into a distillation pot, containing water with hydrosulphite. A part of the methanol distils off into the receiver during the transfer into the pot. The distillation pot and the distillation receiver are closed and the reaction mixture is further heated to 80 to 90 °C until the reaction is terminated. Methanol is then distilled off and the residue in the pot is drained into the isolation vat where it is cooled with brine to 10 °C. The precipitated product is then filtered using a filter press. The material is dried and ground, and after the determination of its colouring strength it is diluted with urea to the necessary concentration.

Optical brightening agents of this type are water-soluble and they give an intense whiteness of a red-violet shade. They are used for the brightening of natural and regenerated cellulose fibres (from weakly acid up to alkaline medium). When in weakly acid medium, they also whiten polyamide, animal fibres and silk. The temperature does not affect their application onto cotton, but it does affect exhaustion of the optical brightening agent onto polyamide and animal fibres. They are used in exhaust and continuous procedures, and also as nuancing components in the dyeing of especially clear and brilliant shades. They can also be added into peroxide and reductive bleaches, into softening and washing baths and into some finishing baths for permanent and non-permanent dressing. In detergents they are especially suitable for cellulose fibres; they can also be added to washing agents, containing enzyme preparations.

4,4′-Bis(6-diethanolamino-4-chloro-1,3,5-triazin-2-yl)aminostilbene-2,2′-disulphonic acid ((17) — compounds of the type Tinopal RP) -

(17)

This brightener is a condensation product of 4,4′-diaminostilbene-2,2′-disulphonic acid, cyanuric chloride and diethanolamine. For its preparation water is introduced into the dissolution vessel and a paste of diaminostilbenedisulphonic acid is suspended in it. Sufficient sodium hydroxide is then added to the suspension to bring the acid into solution as the disodium salt. Ice, emulsifier and cyanuric chloride are introduced into a kettle followed by sufficient hydrochloric acid as to make the reaction acid to Congo Red. After thorough stirring at 0 °C the solution of the disodium salt of diaminostilbenedisulphonic acid, prepared before-hand, is introduced into the kettle. Sodium carbonate solution is then added slowly to keep the mixture acid to Congo Red. When the condensation is complete the mixture is adjusted to pH 6 and it is then transferred into another condensation vessel with simultaneous rapid addition of diethanolamine. The temperature of the reaction mixture is brought to 20 °C and, after thorough stirring, the contents of the vessel are dissolved by heating to 40 − 50 °C. After adjustment of the pH to 6.5 to 7, charcoal is added and the suspension is cleared by filtration into the isolation vessel. The product is salted out with solid sodium chloride and the suspension is transferred through a blow case into a filter press. The paste of the product is put into a mixer, where it is adjusted to the required concentration and dried. The product is suitable for polyamide and animal fibres, to which it lends the appearance of an intense neutral white. It is used either independently in optical brightening, or in other refining processes, as for example reductive whitening or softening with amino active products. It is used in the printing of textiles, either before or after the printing, or may also be added into print pastes.

4,4′-Bis(6-methylamino-4-methoxy-1,3,5-triazin-2-yl)aminostilbene-2,2′-disulphonic acid ((18) — a compound of the type Blankophor PSL)

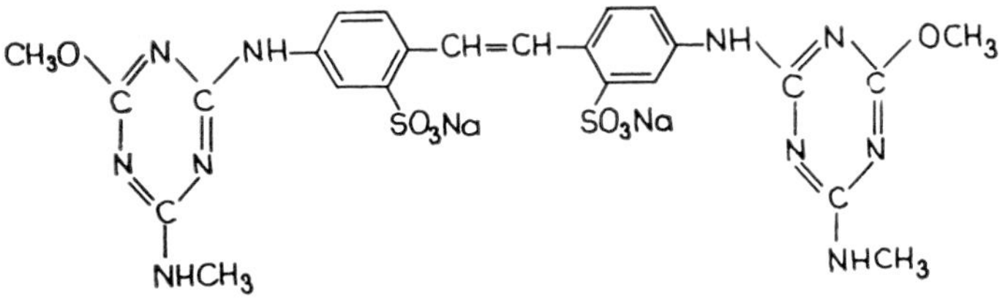

(18)

This product is formed by condensation of 4,4′-diaminostilbene-2,2′-disulphonic acid with cyanuric chloride, methanol and methylamine.

Cyanuric chloride and methanol are introduced into the kettle and stirred at 15 to 16 °C. Ice water is then added to the suspension, followed by sodium hydrogen carbonate. A solution of diaminostilbene-disulphonic acid is then added slowly into the above mixture, which is kept at 30 − 32 °C until a test for free diaminostilbenedisulphonic acid is negative. After the termination of the first condensation, methylamine and sodium hydrogen carbonate are added to the suspension, which is then heated to the boiling point and aqueous methanol is then distilled off. After further refluxing, hydrosulphite and charcoal are added to the distillation residue and the solution is cleared while hot. The filtrate is salted out by addition of sodium chloride. The suspension formed is cooled to 15 °C and filtered. The filter cake is washed with saturated sodium chloride solution and dried. Optical brightening agents of the Blancophor REU type are prepared in the same way, but instead of methylamine alone a mixture of N-methyltaurine and monoethanolamine are used.

4,4′-Bis(6-amino-4-phenylamino-1,3,5-triazin-2-yl)aminostilbene-2,2′-disulphonic acid ((19) — a compound of the Blankophor RA type)

(19)

This compound is formed by condensation of 4,4′-diaminostilbene-2,2′-disulphonic acid with cyanuric chloride, aniline and ammonia.

Water, ice and an aqueous solution of an emulsifier are introduced into the kettle and the contents are acidified with hydrochloric acid. Cyanuric chloride is then added to the mixture with stirring. The suspension is cooled to about 0 °C and an aqueous solution of the sodium salt of diaminostilbenedisulphonic acid is then added slowly at the same temperature. The reaction mixture is then stirred at 2 to 3 °C and the pH is adjusted with a small amount of alkali so that the mixture shows only weak acidity to Congo Red. A pearly shining viscous suspension is formed. After the termination of this condensation, indicated by a negative test for diaminostilbenedisulphonic acid, a solution of aniline hydrochloride is added to the suspension and the reaction mixture is diluted with water. The temperature is increased to 45 °C and kept at this level for 5 to 6 h. The reaction liberates acid which is gradually neutralized with alkali keeping the reaction to Congo Red hardly discernible. The reaction is terminated by stirring at 20 °C overnight. The end of the second condensation step is indicated by the disappearance of aniline from the reaction mixture. The suspension is then heated and the product salted out with sodium chloride. After cooling to 20 °C the product is filtered off. The suction-dried cake is stirred with aqueous ammonia and the paste suspension is heated and brought into solution. It is then introduced as a thin stream with stirring into a mixture of ice, water and sodium chloride. The precipitated yellowish material is filtered off under suction and dried in a vacuum at 60 °C.

In a similar manner a number of other optical brightening agents of this type, called CC/DAS, can be prepared. On the world market about 50 various brands of optical brightening agents of the CC/DAS type are available (21). The most typical are listed in Table 2.

$$(20)$$

The absorption of CC/DAS in the ultraviolet region depends on the substituents, and lies between $\lambda = 347$ nm and $\lambda = 352$ nm. The fluorescence maximum lies between 432 and 442 nm. The substituents in CC/DAS not only affect the absorption or the fluorescence of the compound, but also its application properties. The compounds with an amino group or a secondary or tertiary alkylamino group are suitable for the whitening of paper and cotton at low temperatures, while an anilino group confers affinity for cellulose. The compounds which contain both an aniline group and an amino group are used for laundry detergents, while the compounds with four anilino groups (i.e. $R = R' = -NHC_6H_5$) have the important property that they also brighten polyamide fibres in addition to cellulose. Introduction of a further sulpho group into the aniline residue or the aliphatic imine brings about an improvement in the properties of CC/DAS for the brightening of polyamide and some types of paper; if an additional alkoxy group is introduced into compounds substituted in this way, optical brightening agents are obtained which are fast in acid medium, suitable for the brightening of polyamide fibres and for the simultaneous brightening and finishing of textiles. Optical brighteners of the CC/DAS type are very fast in combined finishing processes with perborates, peroxosulphates, sodium hydrosulphite (dithionite) and sulphite. The most stable ones are those that have morpholine as a substituent, for example compounds of the type Tinopal DMS and BST.

4,4'-Bis(6-phenylamino-4-morpholino-1,3,5-triazin-2-yl)aminostilbene-2,2'-disulphonic acid ((22) — a compound of the Tinopal DMS type)

This is formed on condensation of 4,4'-diaminostilbene-2,2'-disulphonic acid with cyanuric chloride, morpholine and aniline.

Cyanuric chloride is introduced with stirring into a kettle containing cold water, ice and hydrochloric acid, and the suspension is stirred for 15 min. A solution of sodium salt of diaminostilbenedisulphonic acid is then added in a thin stream from a graduated vessel at a speed which keeps the reaction in the kettle slightly alkaline to Congo Red. The mixture is stirred until the diaminostilbenedisulphonic acid has disappeared. After the termination of the first condensation step aniline is poured into the reaction mixture and a solution of sodium hydroxide is added over 1 to 2 hours from a measuring vessel at a speed which does not permit the solution to exceed pH 7. After the addition of sodium hydroxide solution the reaction mixture is stirred until the aniline has disappeared.

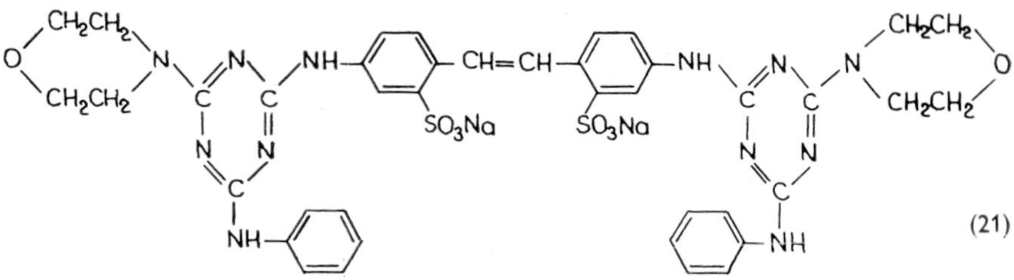

(21)

During stirring the pH of the reaction mixture is constantly checked and it should not exceed the value 7.5. Then the reaction mixture is transferred into the kettle for the second condensation step. Morpholine is added and the mixture heated to 40 °C. A solution of sodium hydroxide is poured slowly into the reaction mixtuie which is heated to 80 °C. The mixture is cooled and thoroughly centrifuged. The paste product is washed (in portions) with cold water and finally with a 10 % sodium chloride solution, until the originally strongly thixotropic paste had assumed a normal consistency. The product obtained is put into a kettle containing diluted ethanol, sodium carbonate, diatomaceous earth and charcoal. The kettle is closed and the reaction mixture is heated at 80 to 90 °C. The solution is pressed over into the crystallization kettle and a filtered, saturated sodium chloride solution is added to it slowly. The mixture is kept boiling until the reaction is over. After cooling, the reaction mixture is centrifuged and washed repeatedly with a mixture of ethanol, water and a saturated sodium chloride solution, until the filtrate is colourless. The centrifuged paste is dried at 80 to 90 °C. Adjustment to the required concentration is effected with sodium carbonate.

Table 2 Survey of commercial brands of optical brightening agents of the CC/DAS type

R^1	R^2	Application	Commercial name
$-Cl$	$-N(CH_2CH_2OH)_2$	polyamide viscose	Tinopal RP
$-NH_2$	$-NHCH_2CH_2OH$	cellulose	Tinopal BU
$-NCH_2CH_2SO_3H$ $\|$ CH_3	$-N(CH_2CH_2OH)_2$	cellulose polyamide paper viscose	Uvitex CK Blankophor REU Ultraphor RPB
$-NH_2$	$-NHC_6H_5$	cotton polyamide	Blankophor RA Albaton 2A, ZN
$-NHEt$	$-NHC_6H_5$	cotton	Uvitex MP, Tinopal BMS
$-NHCH_2CH_2OH$	$-NHC_6H_5$	polyamide	Mikephor BX Uvitex VR
$-N(CH_2CH_2OH)_2$	$-NHC_6H_5$	cotton polyamide	Blankophor BA
$-N(CH_2CH_2OH)_2$	$-NHC_6H_4SO_3H$ (1,3)	viscose in cotton laundry agents	Tinopal 2BF, 2B, 2BN, 2BP, Delft Weiss BNP, Mikephor BN, Stralex 2B, Blankophor BBU
$-N(CH_2CH_2OH)_2$	$-NHC_6H_3(SO_3H)_2$ (1,2,4)	cotton polyamide paper	Blankophor BSU
$-NCH_2CH_2SO_3H$ $\|$ CH_3	$-NHC_6H_4SO_3H$ (1,3)	cotton polyamide paper	Tinopal CH 3690 Mikephor BI
$-NHC_6H_5$	$-NHC_6H_5$	laundry agents cotton polyamide	Blankophor CA 4424, MBBH 766, Uvitex SRBN, SBA, Tinopal TAS, Celamyl DW, Photion D, DK, Delft Weiss NSNW, Stralex 4A, Amar Soap White B, Leukophor BCF, Belofor OD

Table 2 (continued)

R¹	R²	Application	Commercial name
$-NHC_6H_4SO_3H$ (1,4)	$-NHC_6H_4SO_3H$ (1,4)	cotton polyamide	Delft Weiss BSM
$-NHC_6H_5$	morpholino	laundry agents	Tinopal DMS, Mikephor TB, Uvitex SDM, Blankophor MBBH 766
$-NHC_6H_3(SO_3)_2$ (1,2,4)	morpholino	laundry agents	Tinopal BST
$-NHCH_2CH_2SO_3H$	$-NHC_6H_3(SO_3H)_2$ (1,2,4)	laundry agents	Tinopal 3690, BST
$-OCH_3$	$-N(CH_2CH_2OH)_2$	paper	Weisstöner BS, BSP
$-OCH_3$	$-NCH_2CH_2SO_3H$ $\mid$ CH_3	paper	Blankophor PSL
$-OH$	$-NHC_6H_5$	cotton, polyamide, laundry agents, viscose	Blankophor B Uvitex RBS
$-OCH_3$	$-NHC_6H_5$	cotton, polyamide, viscose, laundry agents	Uvitex PRS, CF, SFC, Albaphan 73, CB 180, 2CB 181, Weisstöner BV, Bianco Ottico FC, Blankophor BBH, Fluorolite CN, Whitex BF, Stralex UN
$-NHC_6H_5$	$-NHC_6H_4SO_3H$ (1,3)	cotton, viscose polyamide	Leukophor B

4,4′-Bis[6-(-2,5-disulphophenylamino)-4-morpholino-1,3,5-triazin-2-yl] aminostilbene-2,2′-disulphonic acid ((20) — A compound of the Tinopal BST type)

This product is formed by condensation of 4,4′-diaminostilbene-2,2′-disulphonic acid, cyanuric chloride, 2,5-aminobenzenedisulphonic acid and morpholine.

(22)

Cyanuric chloride and ground limestone are added to a vessel containing water at 10 °C, and a neutral solution of the sodium salt of aminobenzenedisulphonic acid is then added to it slowly. The reaction mixture is kept at 12 to 15 °C until a test for aminobenzenedisulphonic acid is negative. A freshly prepared solution of diaminostilbenedisulphonic acid is added to the vessel and the reaction mixture is stirred, keeping the temperature within 15 to 25 °C. The end of the reaction is determined analytically, and morpholine is added to the reaction mixture which is then heated gradually to 90 to 95 °C, and the third reaction step of the condensation is terminated at this temperature. The reaction mixture is cleared whilst hot, with addition of charcoal and hydrosulphite, and the hot filtrate is salted out in a separate vat with sodium chloride. After cooling with slow stirring down to 20° and then down to 15 °C; the suspension is filtered on a filter press and washed with a salt solution. The paste is suspended in the required amount of water and mixed with anhydrous sodium sulphate and dried.

Optical brighteners of this type show a violet fluorescence and they are poorly soluble in water. They are added to laundry agents. Used in a washing bath they act on cellulose fibres, both natural and regenerated. They show a very favourable accumulation curve on these materials and a shift of the hue takes place only when their concentration on the fibres is considerable. They also increase the whiteness of the laundry agents themselves. They are stable at temperatures used for the production of laundry agents, even in the presence of peroxyborates. They are also suitable for products containing enzymatic preparations, because they are sufficiently active even at relatively low temperatures. They may be combined with other optical brightening agents of both anionactive and non-ionogenic character. They are principally used for anionactive products. Their fastness on cotton is medium to good.

Optical brightening agents derived from stilbene display *cis-trans* isomerism. This property is very important, because both these isomers differ from each other in their colouring and physical properties. Only the *trans* form is coloristically valuable, because it alone fluoresces and possesses the necessary affinity for the substrate. The *cis* isomer does not fluoresce, it is more soluble in water than the *trans* isomer and it does not possess sufficient affinity for the fibre. One form can be converted into the other by irradiation with short-wave length light. The formation of the *cis* form occurs readily in very dilute solutions, and therefore when working with such solutions of *trans*-stilbene the solutions should be protected from light. With concentrated solutions the danger is less, because the surface of the solution acts as a protecting ultraviolet filter.

For application purposes the crystalline form of the product is important, especially in optical brighteners added to laundry powders (containing morpholine, N-methylethanolamine or aniline as substituent). When these substances are produced they are usually obtained as amorphous yellow compounds, and therefore "colour troubles" prevent their combination with colourless, white laundry agents. When brightening agents of this type are boiled at temperatures above 100 °C a so-called white modification is obtained, which is more suitable for combination with laundry agents.

Structurally these CC/DAS compounds include compounds in which the stilbene fluorophore is connected to cyanuric chloride not by means of the imine bridge, but directly by a C—C bond. They are formed by condensation of the chloride of 4,4'-stilbenedicarboxylic acid with benzonitrile in the presence of aluminium chloride, as shown in the reaction sequence (23), or by condensation of 2,4-diphenyl-6-(*p*-tolyl)-1,3,5-triazine with sulphur according to the reaction sequence (24).

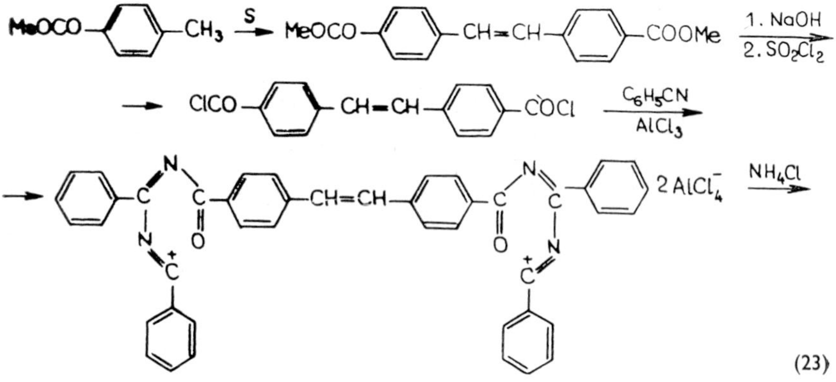

(23)

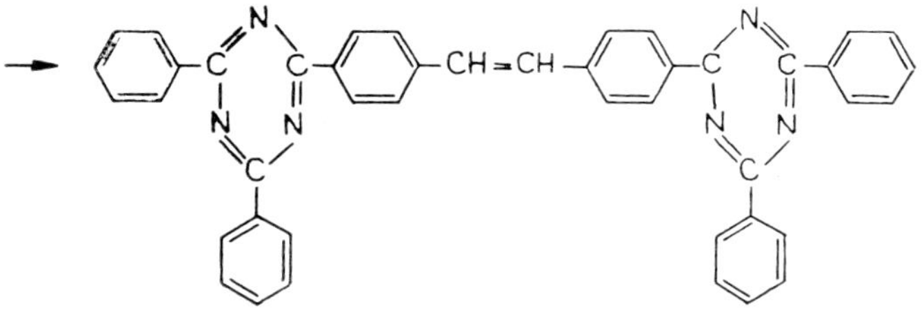

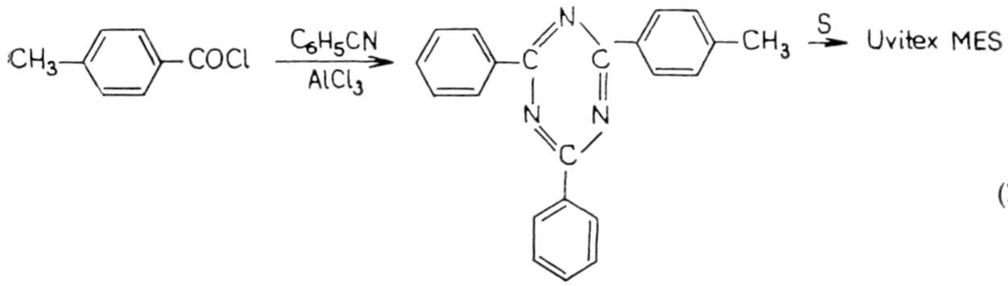

(24)

Uvitex MES is suitable for spunbrightened fibres of polyester, polyamide and polyethylene.

2.1.2 Bisacylamino derivatives of stilbene

Among the acylated derivatives of 4,4'-diaminostilbene-2,2'-disulphonic acid, methoxybenzoyl derivatives of diaminostilbenedisulphonic acid (25) are technically important.

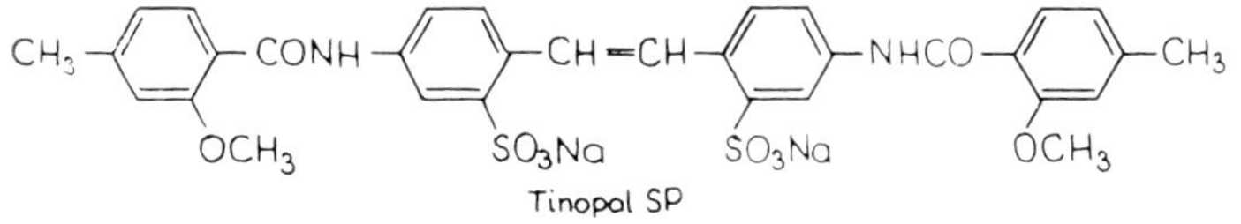

(25)

Tinopal SP

The fluorescent effect increases with the number of methoxy groups. A methoxy group in the 2-position has the most pronounced effect. Optical brightening agents containing methoxy groups in positions 2 and 4 possess optimum brightening properties. Calcofluor White 2GT (26) (Pontamin White 2GT) is an example of such substances.

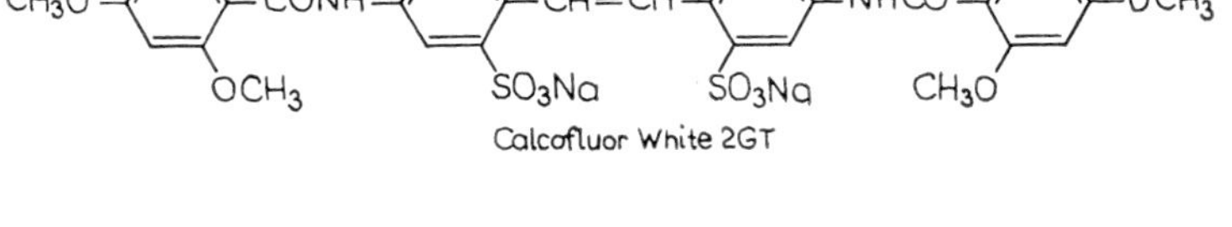

(26)

Calcofluor White 2GT

The introduction of chlorine into positions 5,5′ of the benzoyl substituent causes absorption in the short wave-length region and a reddish fluorescence of the optical brightening agent. In contrast to this, chlorine in the 6,6′ positions decreases the substantivity of the whitening preparation to such extent that it becomes practically negligible.

The 4,4′-diaminostilbene-2,2′-disulphonic acid substituted with chlorine in 5,5′ positions does not give a technically suitable product either, since the fluorescence is shifted into the green region.

Optical brightening agents of the bis-acylamino type are suitable for the brightening of cotton, and polyamides, and also as additives for detergents. They possess excellent affinity for cellulose, but they have poor stability towards chlorinating whitening agents.

2.1.3 Triazole derivatives of stilbene

The triazole derivatives of stilbene may contain one or two triazole residues, in the 4 and 4′-positions. When preparing the monotriazole derivatives, 4-nitro derivatives of stilbene are used as starting material, and these are obtained by Knoevenagel condensation (27) or by Meerwein synthesis (28).

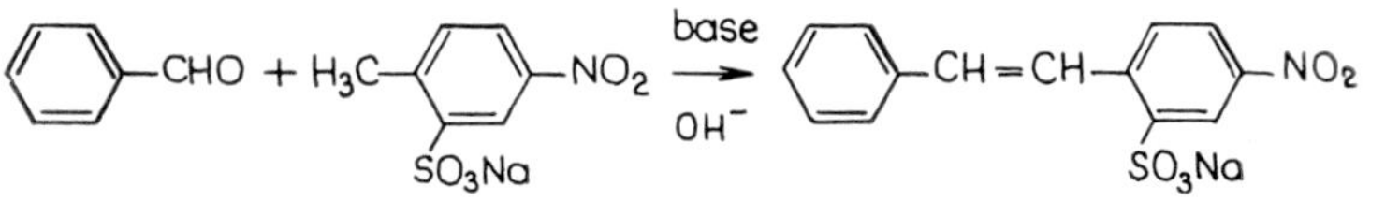

(27)

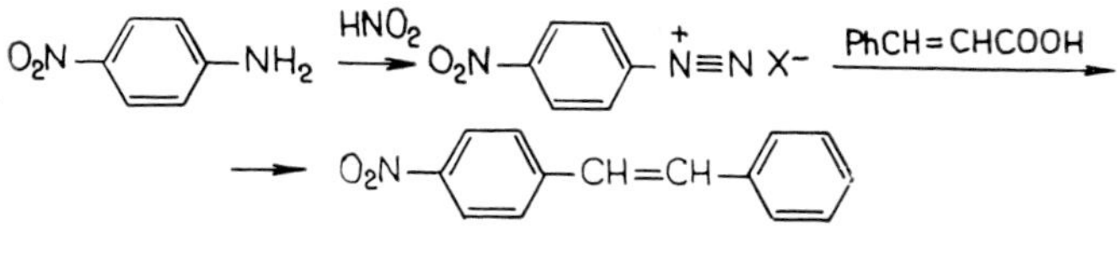

(28)

For example reduction of the nitro group of 4-nitrostilbene-2-sulphonic acid (29-I), gives 4-aminostilbene-2-sulphonic acid (29-II) which is diazotized and coupled with 2-naphthylamine to give the azo dye 29-III. When this o-aminoazo dye is oxidized with ammoniacal copper-II-sulphate, aqueous pyridine and copper-II-sulphate, 12 % sodium hypochlorite solution, air in chlorobenzene or thionyl chloride in pyridine, the triazole ring is formed.

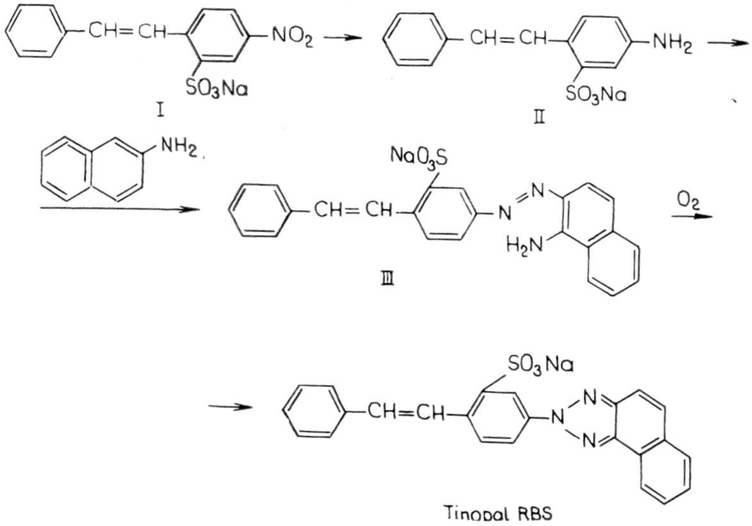

(29)

Tinopal RBS

The end-product is known under the commercial name Tinopal RBS. It gives a neutral fluorescence and exhibits average stability; it is very stable towards hypochlorite and has a good affinity for cotton and polyamides. Tinopal 3511 (30), suitable for the brightening of polyester fibres, and Tinopal PCRP (SFR) (31), used for the brightening of polyvinyl chloride and polyamide in the mass, are prepared in a similar manner.

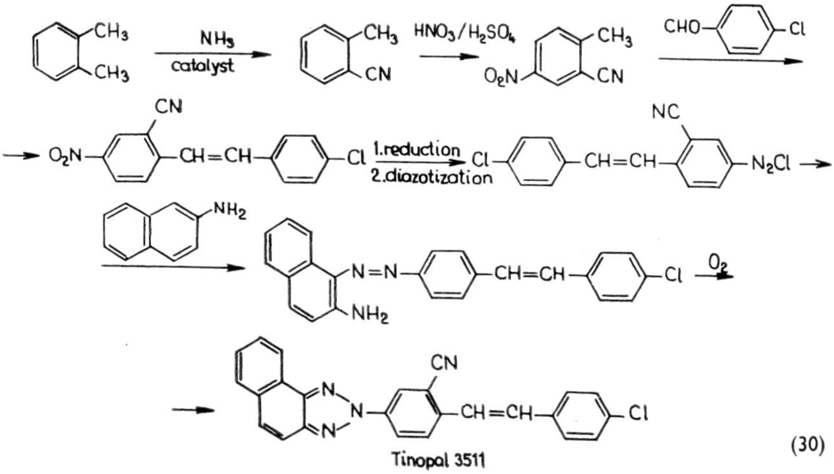

Tinopal 3511

(30)

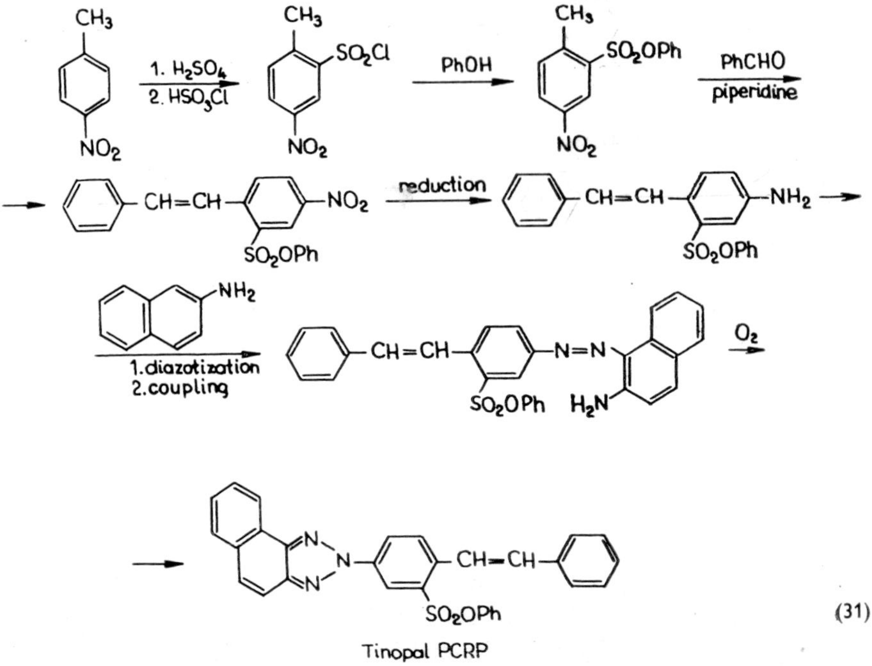

(31)

**4-Chloro-2'-cyano-4'-(naphtho-1,2:3,4-triazol-1-yl)stilbene
(a compound of the Tinopal CH 3511 type)**

Piperidine is added dropwise to a refluxing solution of 4-nitro-2-toluonitrile and p-chlorobenzaldehyde in chlorobenzene and the mixture is refluxed with azeotropic removal of water until the reaction is complete (i.e. water no longer separates). The mixture is allowed to cool and ethanol is then added slowly. The product which crystallizes out after cooling is filtered off under suction and washed in portions with a mixture of ethanol and chlorobenzene and lastly with ethanol. In a reduction kettle iron filings are etched by boiling in water acidified with acetic acid. After this 4-nitro-2-cyano-4'-chlorostilbene suspended in chlorobenzene is added for reduction. When reduction is complete the iron salts are precipitated with sodium carbonate, the precipitate is filtered off and washed with hot chlorobenzene. The latter is steam-distilled off, and the product is filtered off, washed with water and dried. The 4-amino-2-cyano-4'-chlorostilbene formed is dissolved in hot dimethylformamide, a solution of sodium nitrite is added and the mixture poured with rapid stirring into a mixture of hydrochloric acid solution and ice. After further stirring for a short time the excess of sodium nitrite is eliminated with a solution of sulphamic acid and the diazo compound formed is poured slowly into a suspension made by pouring a solution of naphthylamine in pyridine into water. The mixture is stirred for 2 to 3 hours until tests show that the diazo compound has disappeared. The precipitated dye is filtered off under suction, washed with water, dilute hydrochloric acid and again with water. The product is dried and suspended in chlorobenzene, pyridine is added, followed by copper dust and the suspension is heated at 60 to 80 °C. Air is blown through this suspension until decolorization occurs. Acetic acid is then added and the reaction mixture is decolorized with zinc dust. After the addition of ethanol and cooling a precipitate is formed which is filtered off under suction and the cake on the Büchner funnel is washed with hot water. The crude product is suspended in chlorobenzene, and acetic acid and zinc dust are added and boiled until decolorization is complete. After filtration of the suspension through a layer of diatomaceous earth the cake is washed with hot chlorobenzene. Ethanol is added to the filtrate and the solution is cooled to bring about crystallization. After filtering off the precipitate, this was washed and suction dried on a Büchner funnel.

Conversion of Tinopal RBS to its chloride and condensation of the latter with ethylamine or N,N-dimethylethylenediamine gives Tinopal AC (32).

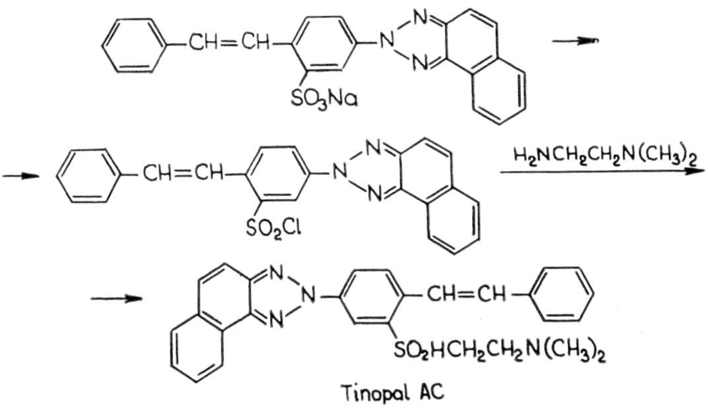

$$(32)$$

Tinopal AC

If 1-naphthylamine-4-sulphonic acid is used instead of 2-naphthylamine for the preparation of the azo dye, preparation (33) is formed,

50

known under the name Mikowhite GS, Tinopal GS or Weisstöner GS with a greenish fluorescence and an affinity for cotton, viscose and polyamide.

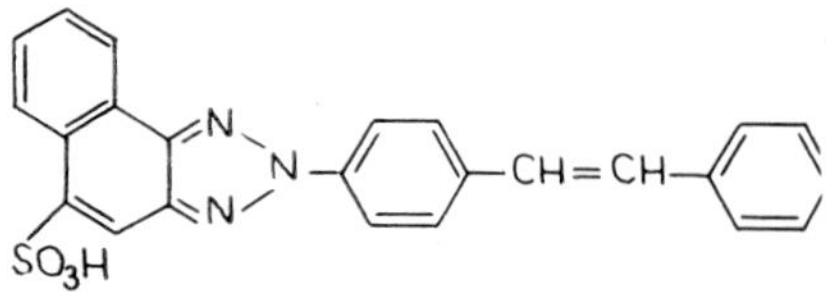

(33)

Using the method described derivatives of stilbene disubstituted with triazole rings can also be prepared. Thus, for example, Blancophor 9 (34) can be formed as follows. 4-4'-Diaminostilbene-2,2'-disulphonic acid is tetrazotized, coupled with 2-naphthylamine-5-sulphonic acid and the azo dye formed is cyclized by oxidation.

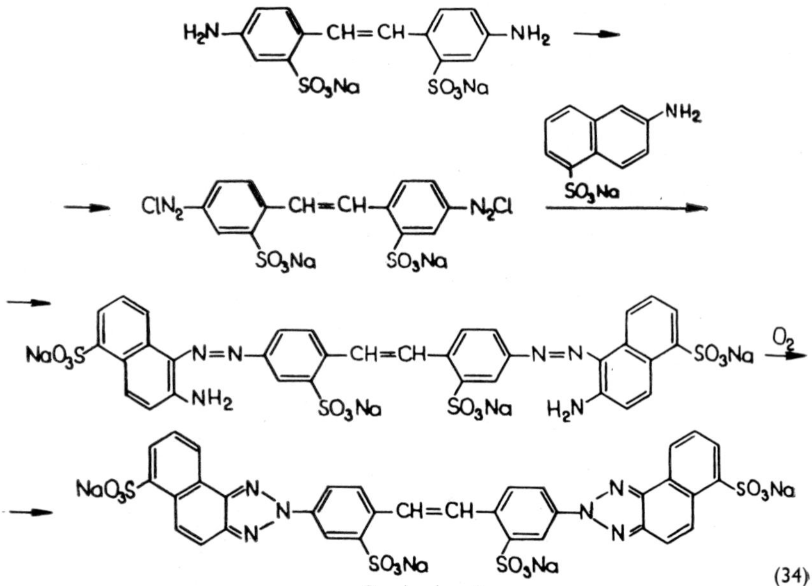

(34)

Similar substances (35) can also be prepared by oxidation of 2-p-tolylbenzotriazole in the presence of sulphur.

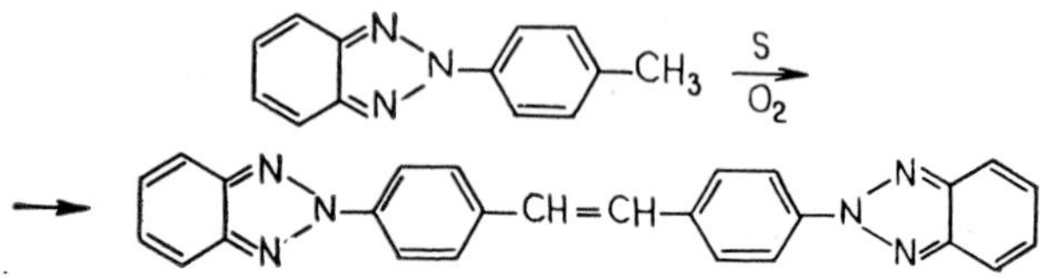

(35)

In addition to the above mentioned industrially produced products a number of similar substances have been proposed as suitable optical whitening agents.

The firm Ciba–Geigy has patented optical brightening agents prepared by the Siegrist anil synthesis from tolylnaphthotriazole and the anil of *p*-methoxybenzaldehyde (36).

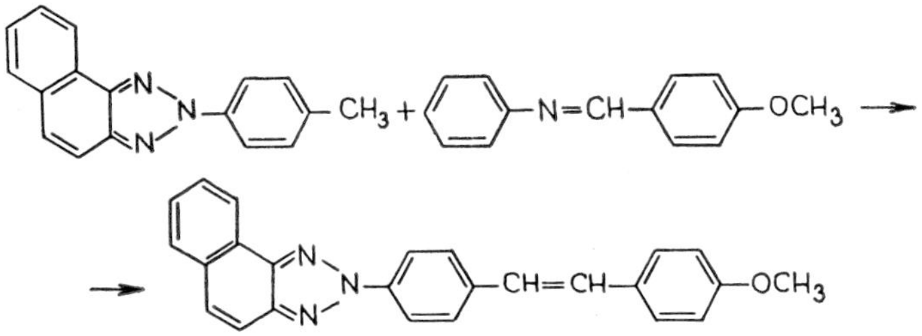

$$\text{(naphthotriazole)}-\text{CH}_3 + \langle\text{Ph}\rangle-\text{N}=\text{CH}-\langle\text{Ph}\rangle-\text{OCH}_3 \longrightarrow \tag{36}$$

$$\longrightarrow \text{(naphthotriazole)}-\langle\text{Ph}\rangle-\text{CH}=\text{CH}-\langle\text{Ph}\rangle-\text{OCH}_3$$

In the same manner the Japanese firm Nippon Kagaku Co. Ltd. prepares the sulphonated derivative (37).

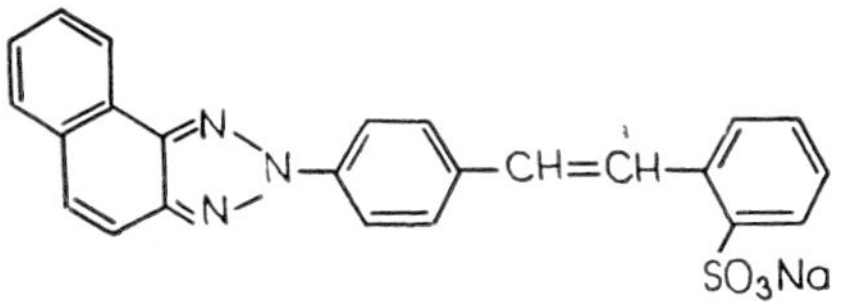

$$\text{(naphthotriazole)}-\langle\text{Ph}\rangle-\text{CH}=\text{CH}-\langle\text{Ph(SO}_3\text{Na)}\rangle \tag{37}$$

The GAF Corp. have proposed the preparation of optical brightening agents with substituted triazoles (38).

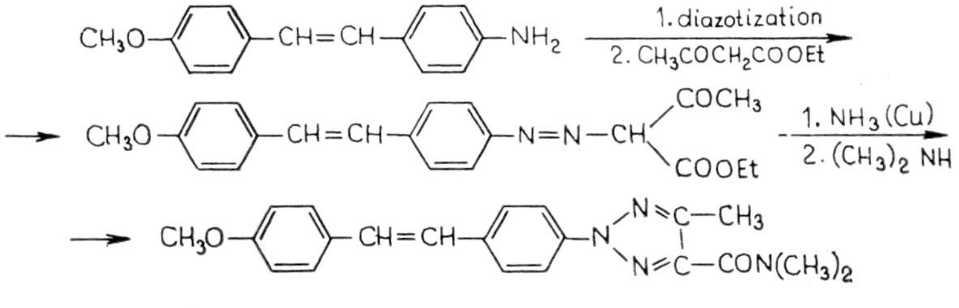

$$\text{CH}_3\text{O}-\langle\text{Ph}\rangle-\text{CH}=\text{CH}-\langle\text{Ph}\rangle-\text{NH}_2 \xrightarrow[\text{2. CH}_3\text{COCH}_2\text{COOEt}]{\text{1. diazotization}}$$

$$\longrightarrow \text{CH}_3\text{O}-\langle\text{Ph}\rangle-\text{CH}=\text{CH}-\langle\text{Ph}\rangle-\text{N}=\text{N}-\text{CH}\overset{\text{COCH}_3}{\underset{\text{COOEt}}{}} \xrightarrow[\text{2. (CH}_3)_2\text{NH}]{\text{1. NH}_3\text{(Cu)}} \tag{38}$$

$$\longrightarrow \text{CH}_3\text{O}-\langle\text{Ph}\rangle-\text{CH}=\text{CH}-\langle\text{Ph}\rangle-\text{N}\overset{\text{N}=\text{C}-\text{CH}_3}{\underset{\text{N}=\text{C}-\text{CON(CH}_3)_2}{}}$$

The Companies of Bayer and Ciba have described the simple bis(benzotriazolyl)stilbene derivative (39).

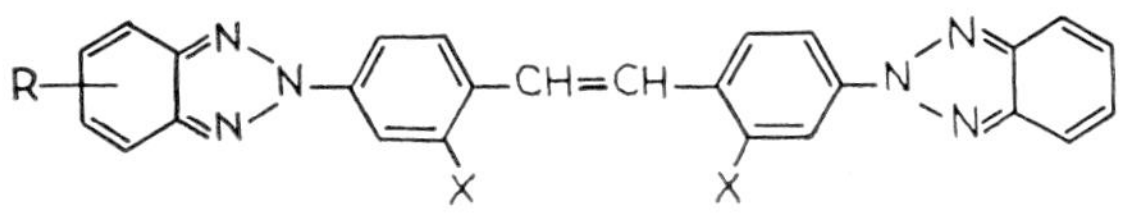

$$\text{R}-\text{(benzotriazole)}-\langle\text{Ph(X)}\rangle-\text{CH}=\text{CH}-\langle\text{Ph(X)}\rangle-\text{(benzotriazole)} \tag{39}$$

52

In addition to vicinal 1,2,3-triazoles, 1,3,4-triazoles have also been described as optical brightening agents, prepared according to scheme (40).

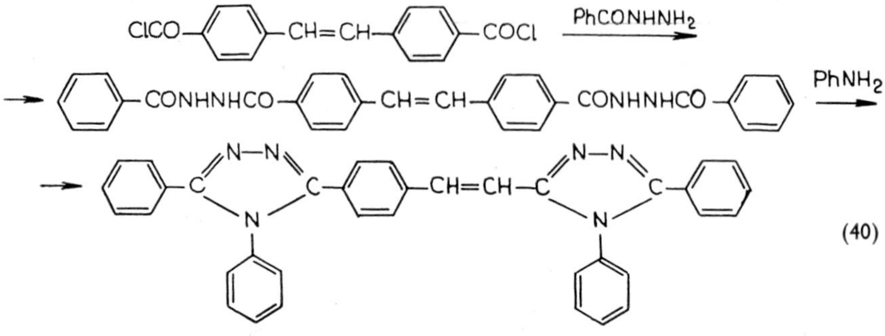

(40)

Optical brighteners suitable for cotton and polyamides have also been prepared by more complex procedures, so far industrially unfeasible-such as condensation of 4-azido-4'-nitrostilbene-2,2'-disulphonic acid with acetylene (41), condensation of dihydrazinostilbene with nitroacetal, dehyde oxime (42), and condensation of dihydrazinostilbene with diacetyl monoxime (43) or with isonitrosoacetophenone (44).

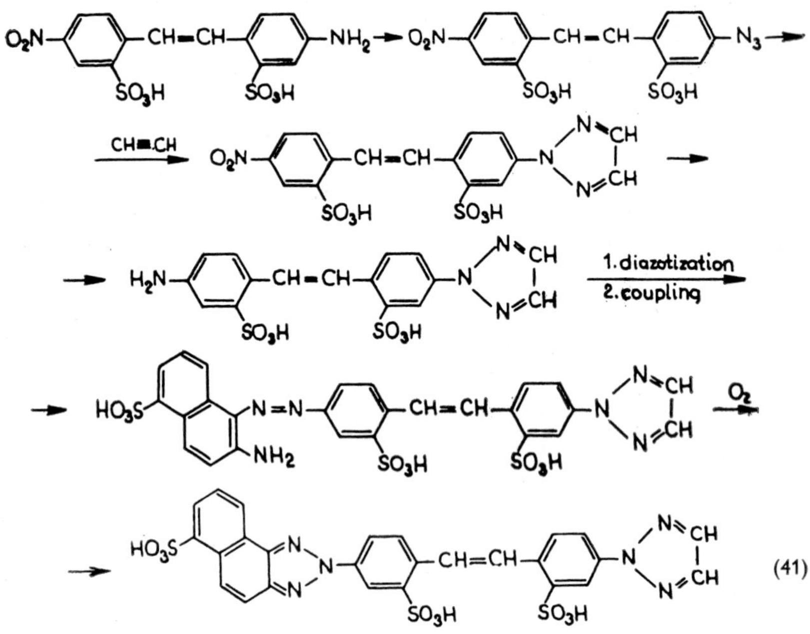

(41)

Triethyl phosphite is condensed directly with the derivatives of benzyl chloride without a condensation agent.

In the case of dimethyl phosphite sodium methoxide has to be used during condensation, forming first the sodium salt of dimethyl phosphite which is then condensed with the corresponding benzyl chloride. The phosphate formed is condensed with terephthalaldehyde in the presence of KOH in dimethylformamide. In this manner the important optical whitening agent for polyester is formed, called Palanilbrillantweiss R and Uvitex OB, during the preparation of which diethyl 2-cyanobenzyl-phosphate and terephthalaldehyde, or 1,4-xylenediphosphonate and 2-cyanobenzaldehyde are used as starting compounds. The preparation of diethyl 2-cyanobenzylphosphonate from triethyl phosphite takes place according to reaction (56), from dimethyl phosphite according to reaction (57) and from 1,4-bis-(chloromethyl)benzene according to reaction (58).

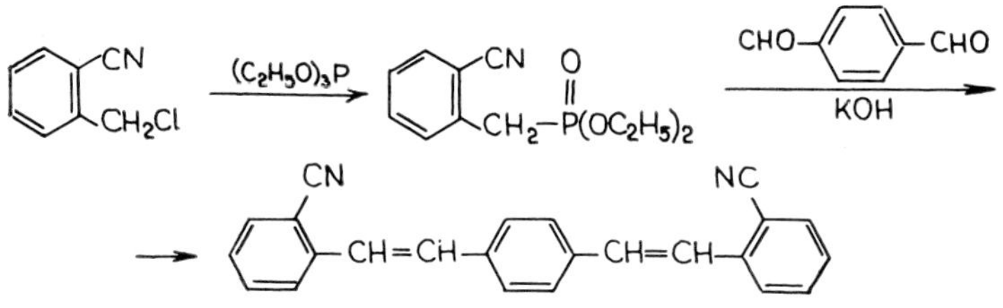

Palanilbriliantweiss R

(56)

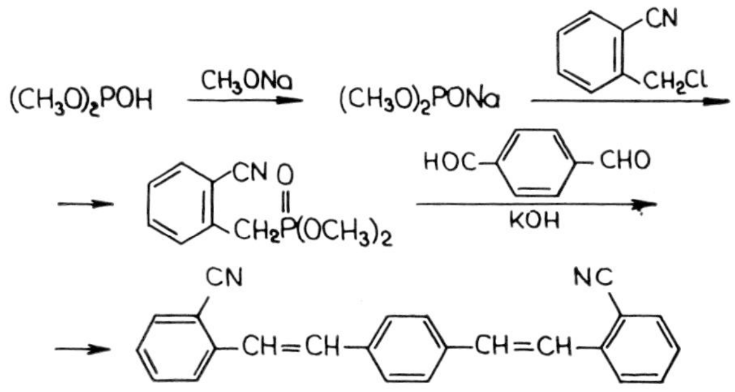

(57)

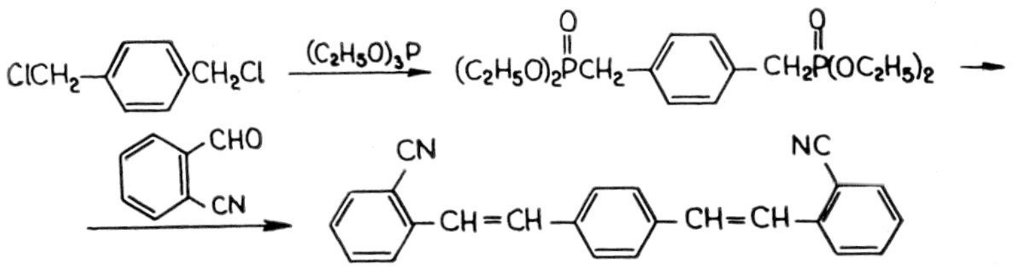

(58)

The preparation of Uvitex 3257 (59) from xylenyl chloride and terephthalaldehyde is also based on Wittig's synthesis.

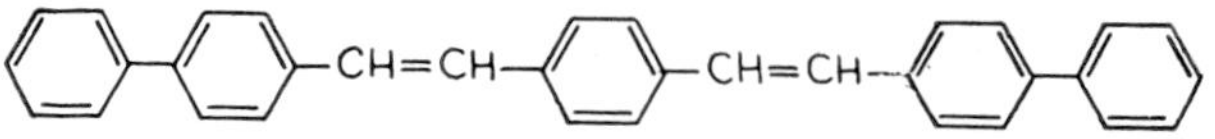

(59)

To illustrate the procedure for the preparation of a substance of the Palanilbrillantweiss R type, both reactions involving triethyl phosphite and dimethyl phosphite are described below.

1,4-(2′,2′-cyanostyryl)benzene
(a compound of the Palanilbrillantweiss R type)

Preparation with triethyl phosphite

A mixture of 2-cyanobenzyl chloride and triethyl phosphite in a kettle provided with a reflux condenser is heated at 150 °C until the reaction is complete. The excess of triethyl phosphite is distilled off and the phosphonate formed is used without isolation for the final condensation. This is effected by addition of a solution of terephthalaldehyde in dimethylformamide and an ethanolic solution of sodium hydroxide. The solution is heated at 60 °C until the reaction has ceased, then cooled to 20 °C and the crystalline material is filtered off on a suction filter. The cake is washed with ethanol, triturated with water and filtered off again, washed with water and suction-dried.

Preparation from dimethyl phosphite

2-Cyanobenzyl chloride, dimethyl phosphite and benzene are introduced into a kettle provided with a reflux condenser. The mixture is mildly heated until homogeneous and then cooled to 40 °C. At this temperature the sodium methoxide solution in methanol is added slowly and the reaction mixture refluxed for a short time. The solvent is then distilled off and when the distillation residue has cooled to 20 °C water is added to the remaining suspension, which dissolves the sodium chloride separated during the reaction. The suspension is transferred into a separatory funnel and the bottom layer, representing crude oily cyanobenzyl dimethyl phosphonate, is separated. The upper, aqueous layer is extracted with benzene. The benzene extracts are combined with the main phosphonate fraction, which is then dried over anhydrous sodium sulphate and the benzene distilled off. Ethanol is added to the residue, the suspension formed is brought into solution by heating, and the solution is cooled with brine. The product which crystallizes out is filtered off under suction and washed with a small amount of ethanol and dried. A methanolic solution of potassium hydroxide is added at 20 °C into a mixture of the so-formed phosphonate, terephthalaldehyde and dimethylformamide. Immediately a green-yellow precipitate is formed and the temperature rises spontaneously to 40°C. After the addition of the methanolic solution of KOH the mixture is heated at 60 °C until the reaction is over. After cooling to 20 °C the solution is poured into water and the separated substance is filtered off under suction, washed with 50 % ethanol and dried.

2.2 Derivatives of pyrazoline

The discoverer of this class of optical brightening agents was Knorr, who examined the reduction of pyrazoles, obtained from 1,3-di-carboxylic compounds and hydrazine. He thus prepared pyrazoline derivatives, which displayed blue fluorescence.

A general method for the preparation of optical brighteners of the pyrazoline type involves the condensation of α, β-unsaturated carbonyl compounds with phenylhydrazine (60).

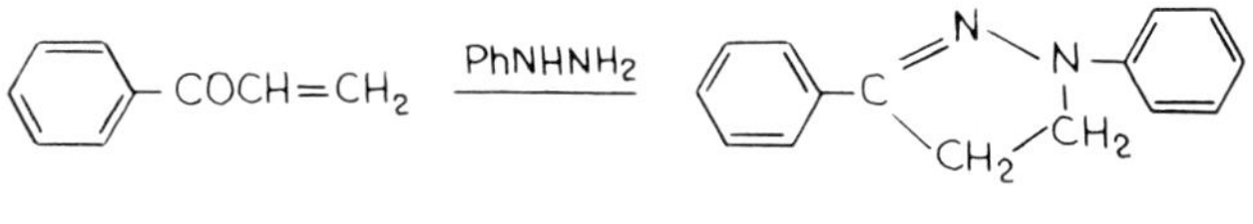

(60)

Condensation of phenylhydrazine with derivatives of phenyl vinyl ketone does not always take place well, because the vinyl compounds are poorly stable, and therefore this reaction is best carried out with tri(β-benzoylethyl)amine hydrochloride, which is prepared by Tollens reaction from acetophenone, formaldehyde and ammonium chloride (61).

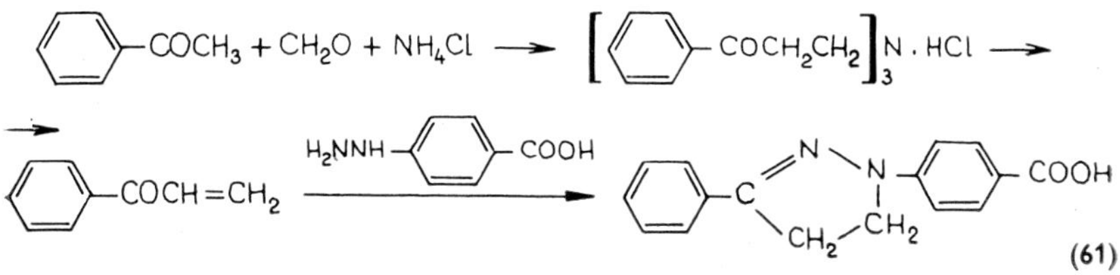

(61)

Blankophor DCB or Blankophor FB 766 are probably prepared by this method.

An analogy to this procedure is the condensation of ketones with formaldehyde and dimethylamine, giving Mannich bases (62). Neither reaction is particularly elegant in spite of their general validity and they give poor yields.

A further method of preparation is the condensation of an aryl-hydrazine with β-chloropropiophenone or its derivatives and the cyclization of the hydrazone formed to a pyrazoline derivative (63).

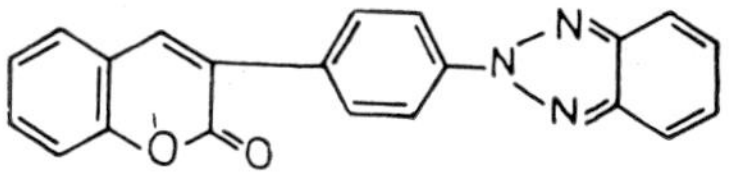

(62)

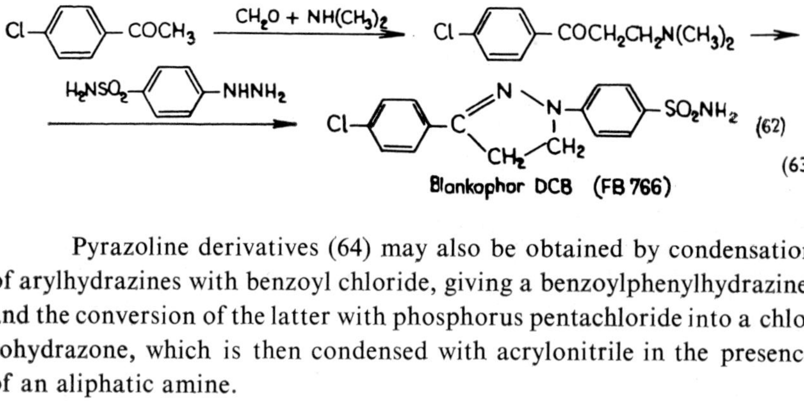

(62)

(63)

Blankophor DCB   (FB 766)

Pyrazoline derivatives (64) may also be obtained by condensation of arylhydrazines with benzoyl chloride, giving a benzoylphenylhydrazine, and the conversion of the latter with phosphorus pentachloride into a chlorohydrazone, which is then condensed with acrylonitrile in the presence of an aliphatic amine.

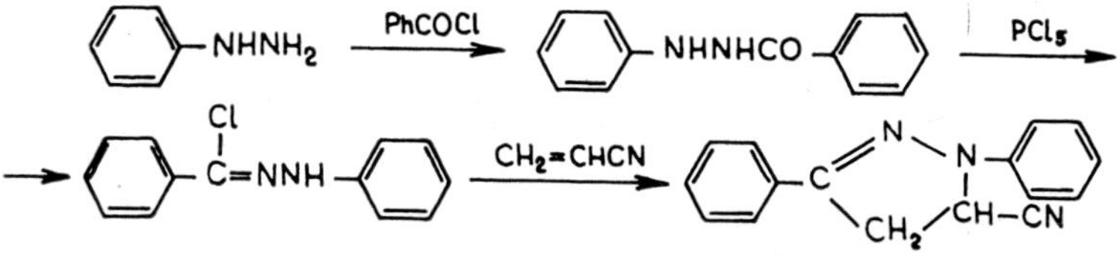

(64)

In addition to the already mentioned industrially important procedures, the derivatives of 1,3-diphenylpyrazoline may also be obtained by arylation of 1-phenyl-Δ^2-pyrazoline with a diazoni um compound in position 3, according to scheme (65), or by dipolar 1,3-addition of com-

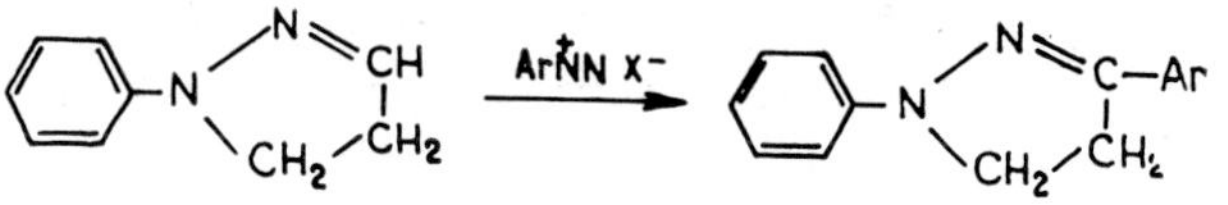

(65)

pounds of the type $R-CH=N-NH-R$ (phenylhydrazones of aromatic aldehydes) to olefins (66).

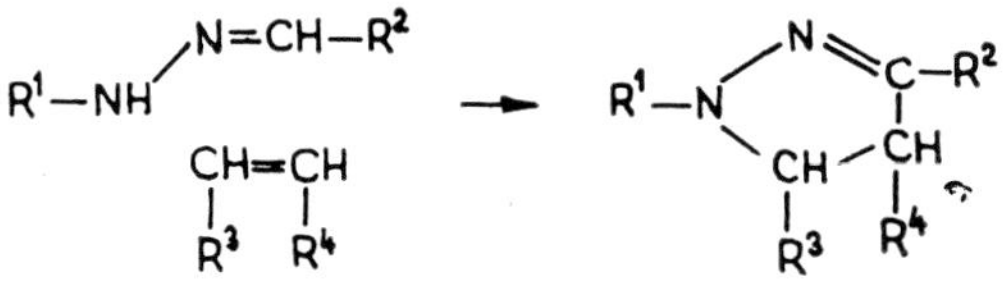

(66)

These reactions are suitable for the introduction of a specific substituent into the 3-position of the phenyl ring, or into the 5-position of the pyrazoline skeleton. They are of no industrial importance.

In addition to the relatively simple pyrazoline compounds mentioned, a series of optical brightening agents of the pyrazoline type has been described where the diphenylpyrazoline skeleton proper is condensed with some other aliphatic or aromatic system (67).

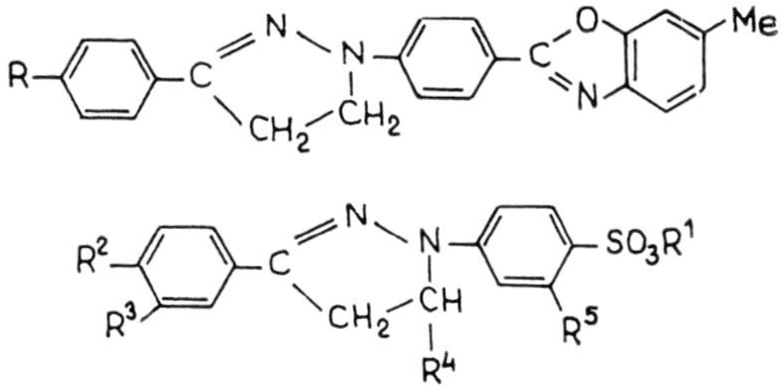

(67)

Pyrazoline optical brightening agents display intensive blue fluorescence. Only derivatives of 1,3-diphenylpyrazoline fluoresce strongly. Substances with absorption about 390 nm are most convenient. Derivatives of 1-phenyl-3-alkylpyrazoline fluoresce only weakly or not at all. The majority of commercial brands are derivatives of 1-phenyl-3-(p-chloro-phenyl)-pyrazoline (Tab. 3). The sulphamide group introduced in the para-position of the phenyl nucleus at position 3 of the pyrazoline skeleton lends to substances of this type excellent application properties, especially a high affinity and substantivity for the fibres, and high brilliance and brightening effects. Optical brightening agents based on pyrazoline are used mainly for the surface brightening of polyamide, acetate and polyacrylonitrile, and as an addition into papers and saponate preparations. They are unstable against oxidants. On reaction with lead tetra-acetate, a primary product of oxidation is formed, i.e. the corresponding pyrazole. Other oxidants, such as mercuric oxide, mercuric acetate, bromine, chromi-

Table 3   Commercially produced optical brightening agents of the pyrazoline series

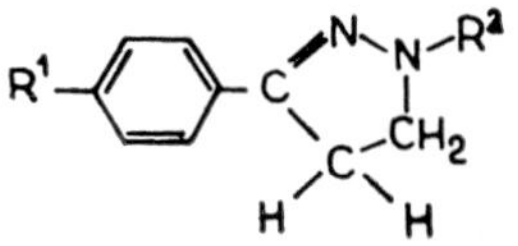

Commercial name	R^1	R^2
Tinopal CH 3584	—H	—Cl
Tinopal WG	—H	—SO$_3$Na
Blankophor CA 4236	—Cl	—SO$_3$CH$_3$
Blankophor DCR flüss.	—Cl	—SO$_3$CH$_3$
Blankophor DCB, FB 766, FD	—Cl	—SO$_2$NH$_2$
Hostalux PR	—H	—SO$_2$CH$_2$CH$_2$OSO$_3$Na
Blankophor DCB 80	—Cl	—SO$_2$CH$_2$CH$_2$NH$_2$

um trioxide, silver nitrate and potassium permanganate oxidize pyrazoline products to complex mixtures.

2.3   Derivatives of coumarin

In 1929 Krais tested for the first time the whitening effect of esculin (69), the spectral properties of which had already been described by Kopp and Joseph in 1911. Hydroxycoumarin itself is not adequately substantive, but good optical brightening agents have been obtained from amino derivatives of coumarin. Efficient coumarin optical brightening agents have their fluorescence maximum at $\lambda = 400$ to 500 nm, and those industrially most interesting are based on derivatives of 3-phenyl-7-amino-coumarin. Coumarin derivatives used today are effective for animal textile materials and for synthetic fibres, for example polyamide and acetate fibres. They are used for the whitening of rabbit furs and feathers, they are added to photographic emulsions (as antifogging agents) to soap powders, cosmetic creams, (as protective agents against sunburn and are used for the whitening of foods (they are not harmful to health).

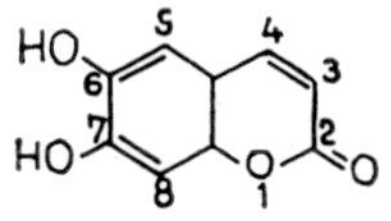

(69)

2.3.1 Coumarin derivatives substituted in the 3-position

3-Carboxy derivatives of coumarin (70) represent a not especially important group of optical brighteners, obtained by condensation of 2-hydroxy-1-naphthaldehyde with malonic ester in the presence of pyridine or morpholine.

Optical brightening agents of this type have average colouring properties and they are suitable for the whitening of polyesters, polyurethanes, polyacrylonitrile, polyamides and viscose. It is said that the amides have better properties than esters.

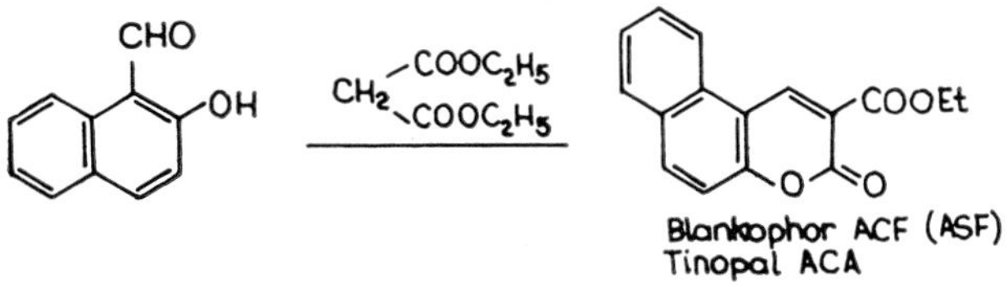

Blankophor ACF (ASF)
Tinopal ACA

(70)

Derivatives of 3-phenylcoumarin are industrially important. 3-Phenylcoumarin is synthesized by Knoevenagel reaction of salicylaldehyde and benzyl cyanide or nitrobenzyl cyanide (71); the reaction with o-methoxybenzaldehyde is smoother.

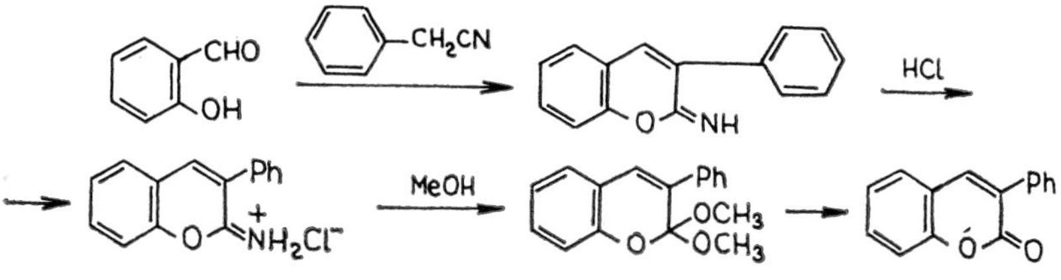

(71)

A brightener with a triazole substituent in the phenyl nucleus, prepared by this synthesis or by the Perkin–Oglialorov or Meerwein procedure described below, is being put on the market under the trade name Leukophor EGU (72).

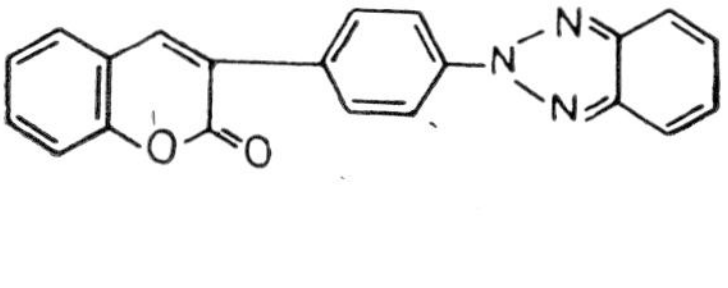

(72)

The brightener (73) with a condensed benzene ring has similar colouring properties.

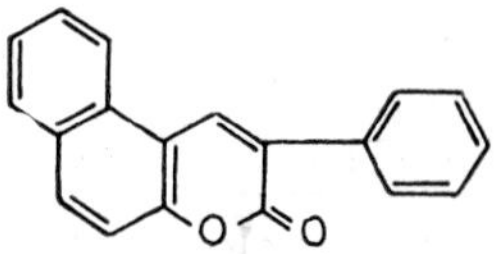

(73)

The Perkin—Oglialorov method of condensation involves the reaction of salicylaldehyde with phenylacetic acid in acetic anhydride or sodium acetate (74).

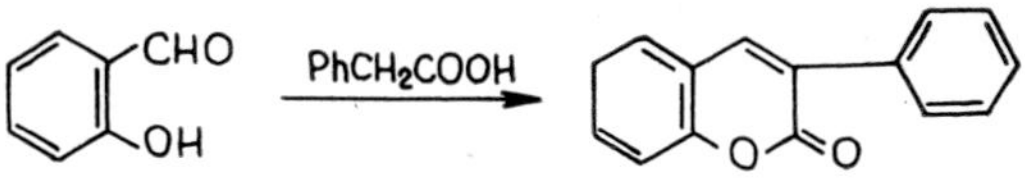

(74)

Another method of preparation is the Meerwein arylation of α, β-unsaturated aldehydes or ketones (75) and the cyclization of 2-nitro-α-phenylcinnamic acid (76).

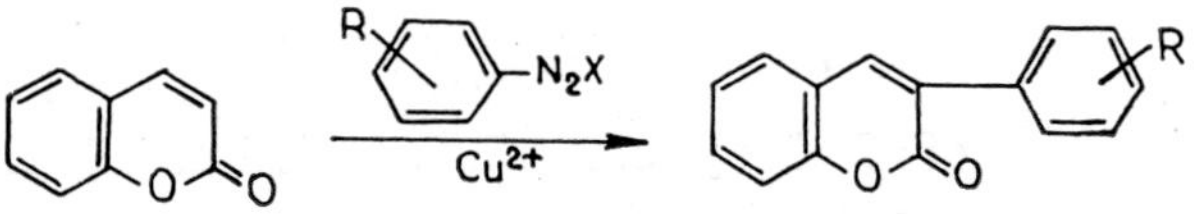

(75)

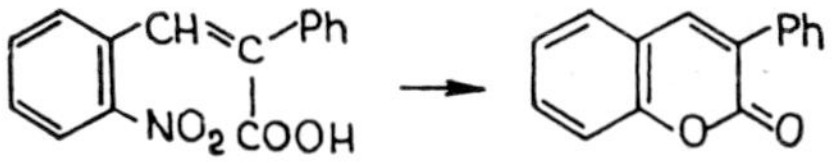

(76)

2.3.2 Coumarin derivatives substituted in the 7-position

Among derivatives of 7-aminocoumarin, 7-diethylamino-4-methyl-coumarin (77) is industrially important, which is prepared from *m*-diethyl-aminophenol and ethyl acetoacetate in the presence of zinc chloride.

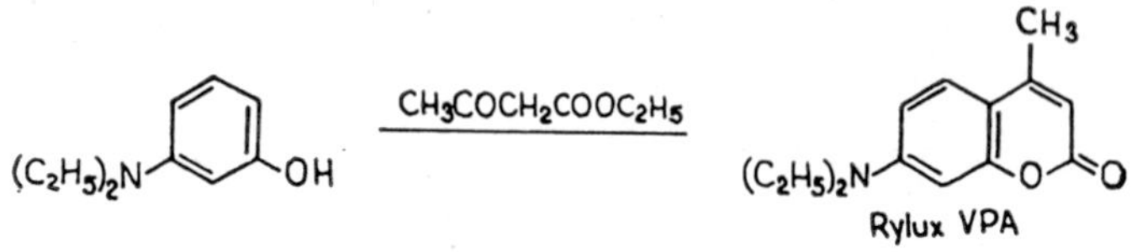

(77)

The product obtained is marketed under various commercial names (Blankophor FBO, Delft Weiss LS, Leukophor WC, Leucophor DC,

Uvitex SWN, Uvitex SWR) and is used for the brightening of paper, cellulose acetate, polyamides, polyacrylonitrile and polyester.

7-Dimethylamino-4-methylcoumarin
(a compound of the Uvitex SWN type)

This is prepared from N,N-diethyl-m-aminophenol by condensation with ethyl acetoacetate.

Preparation of N,N-diethyl-m-aminophenol

A reaction kettle is filled first with a concentrated potassium hydroxide solution and the contents are heated to 250 °C, when a diluted paste of diethylmetanilic acid is added to the melt gradually with stirring. [The acid is obtained by ethylation of metanilic acid with ethyl chloride in alkaline medium in an autoclave, followed by salting out with an excess of sodium hydroxide.] After addition of all the diethylmetanilic acid the melt is heated at 260 to 270 °C until the reaction is complete. The melt thickens gradually, but still remains fluid. Water is then added carefully to the melt (in order to prevent boiling over). The contents of the kettle are transferred by pressure into a vat containing water and ice, and after several minutes' stirring the undissolved potassium sulphite is filtered off under suction. The sulphite paste on the Büchner funnel is washed with a saturated sodium sulphite solution. The combined strongly alkaline filtrates are neutralized with concentrated hydrochloric acid until they give a weakly alkaline reaction. The crude diethyl-m-aminophenol separates from the solution in the form of coarse grey-brown crystals. The semicrystalline mixture is cooled by addition of ice and then allowed to stand overnight and filtered off under suction. The paste on the Büchner funnel is washed with water, suction-dried and then vacuum-dried in a drying oven. The crude diethyl-m-aminophenol is distilled in a vacuum under nitrogen. The main fraction distils between 164 and 165 °C at 1.733 kPa ($= 13$ Torr).

Condensation

Equimolar amounts of ethyl acetoacetate and diethyl-m-amino-phenol are introduced into a kettle with a reflux condenser; the mixture is heated to 60 °C and zinc chloride is added at this temperature. The reaction mixture is refluxed until the reaction is complete and the ethanol formed in the reaction is then distilled off.

The condensation solution in the pot is allowed to cool and is acidified with acetic acid and transferred into the isolation vat containing a mixture of water and ice. The product first separates as an oil, which eventually solidifies to a coarse crystalline material. The suspension is transferred into a Büchner funnel for filtration, washed with water and dried in a vacuum drying-oven. The product is used for the brightening of wool, natural silk, polyamides, acetate, cellulose triacetate and polyacrylonitrile. It is also added to washing powders. It has a reddish shade and is effective within the pH 3 to 11 range. It is insoluble in water, and soluble in alcohols and glycols. In the textile industry 0.05 to 0.3 % of it is added for polyamide, 0.05 to 0.35 % for wool, 0.05 to 1.40 % for cellulose acetate, 0.10 to 2 % for cellulose triacetate, and 0.05 to 0.8 % for polyacrylonitrile.

2.3.3 Coumarin derivatives substituted in the 3- and 7-positions

From the application point of view, derivatives of 3-phenyl-7-aminocoumarin are most interesting. They are prepared similarly to the derivatives of 3-phenylcoumarin, using the Knoevenagel or Perkin—Oglialorov synthesis or Meerwein's arylation.

3-Phenyl-7-aminocoumarin (78) is formed on condensation of m-aminophenol with ethyl 2-hydroxy-1-phenylacrylate or 3-hydroxy-2-phenyl-acrylonitrile.

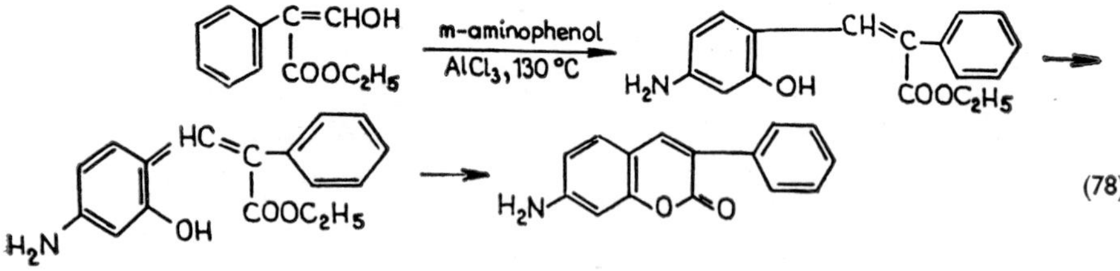

(78)

3-Phenyl-7-aminocoumarin is the starting material for a number of coloristically interesting optical brightening agents. On acylation with cyanuric chloride N-triazinyl derivatives are formed (for example 79).

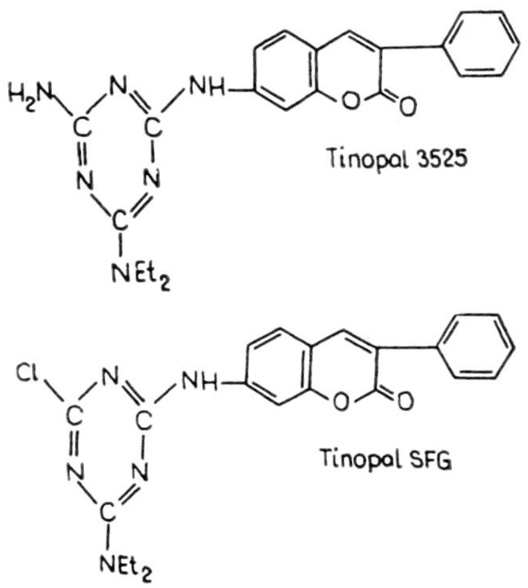

Tinopal 3525

(79)

Tinopal SFG

On diazotization of 3-phenyl-7-aminocoumarin, coupling to 2-naphthylamine and subsequent oxidation, the triazole derivative Leukophor EFG (80) is formed which is suitable for polyester and plastics.

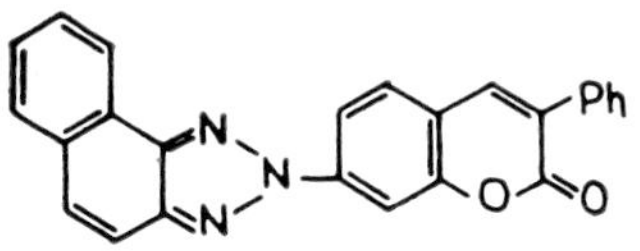

(80)

The analogous reaction (81) involving 5-amino-2-butoxytoluene takes place in a similar manner.

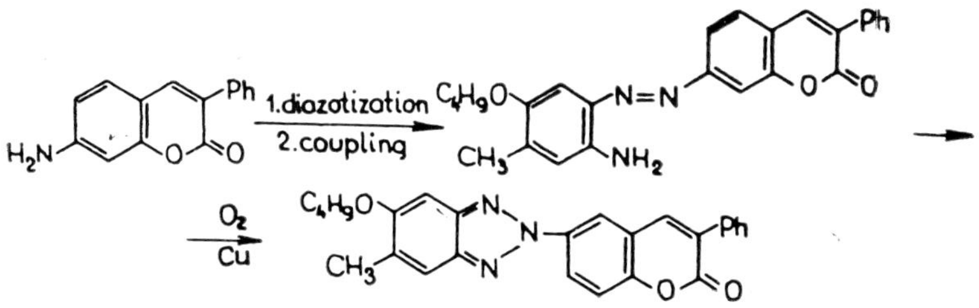

(81)

Reduction of the diazonium compound of aminocoumarin gives rise to 3-phenyl-7-hydrazinocoumarin which is condensed with 1,3-diketones to afford derivatives of pyrazolecoumarin; for example, on condensation with 3-oxobutanal dimethylacetal 3-phenyl-7-(3'-methyl-pyrazol-1'-yl) coumarin (82-I) is formed, and 3-phenyl-7-(3',5'-dimethyl-pyrazol-1'-yl) coumarin (82-II) with acetylacetone. Both products are suitable for the brightening of polyester, cellulose acetate, polyamide 6 and 66.

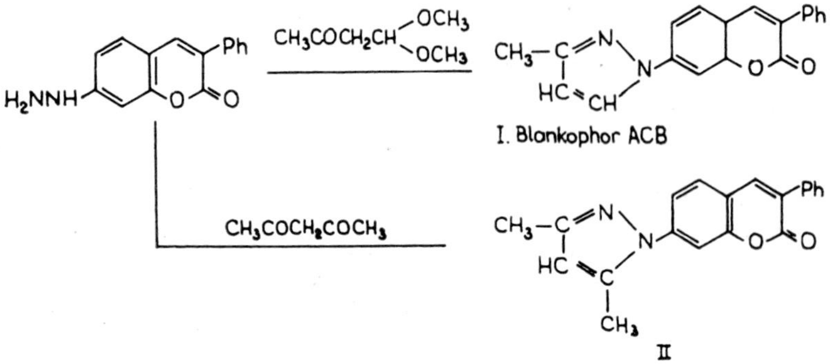

I. Blonkophor ACB

(82)

II

**3-Phenyl-7-(3',5'-dimethylpyrazol-1'-yl)coumarin
((82) — a compound of the Blankophor ACB type)**

Preparation of methyl phenylformylacetate

An ethanolic solution of sodium ethoxide is admitted into a duplicator containing toluene, and the azeotropic mixture of ethanol-toluene-water is distilled off until the temperature in the column head has attained 106 to 110 °C. The duplicator is then cooled and a mixture of methyl phenylacetate and anhydrous ethyl formate is added. After 5 minutes' stirring at 50 °C the reaction mixture in the duplicator is cooled and let out into a mixture of cold water and ice in a vat provided with a stirrer. After short stirring the suspension is allowed to settle and the lower layer is transferred into a container. The toluene remaining in the vat is re-extracted with water and the aqueous layer is added to the container with the main fraction. The toluene is submitted to recovery. The aqueous fractions are retransferred into the washed vat, ice and chloroform are added and the mixture is acidified with sulphuric acid (diluted by pouring the concentrated acid onto ice). After short stirring the suspension is allowed to settle and the lower, chloroform layer is drawn off into a container. The remaining aqueous layer is reextracted with chloroform and the chloroform extract is combined with the main fraction. The aqueous fraction in the vat is drained off, the vat is rinsed with water and the chloroform extract is returned into it. There it is purified by stirring with cold water and the settled chloroform layer is drawn off into the container. The residue in the vat is again extracted with chloroform and the chloroform layer is added to the remaining fraction of the chloroform extract. This is transferred into a blow case, where it is dried over anhydrous sodium sulphate and filtered through a lens filter into the distillation pot. Chloroform is then distilled off initially under atmospheric pressure and finally in a vacuum. The residue in the kettle is discharged into a reservoir.

The crude methyl 3-hydroxy-2-phenylacrylate obtained is further processed without purification.

Preparation of 3-phenyl-7-aminocoumarin

o-Dichlorobenzene and m-aminophenol are introduced into a duplicator and the mixture is heated to 175 °C with stirring. The corresponding amount of crude methyl phenylformylacetate from the preceding preparation is added to the mixture and a mixture of dichlorobenzene, water and methanol is simultaneously distilled off. When all water is distilled off the distillation is switched over to refluxing which is continued for 3 h. The contents of the kettle are then cooled and rapidly diluted with ethanol. The suspension is filtered off on a suction filter and suction-dried. The final filtrate should be clear and the product on the filter a clear yellow. The product is washed with cold water, submitted to strong suction and dried in a drying-oven.

Preparation of 3-phenyl-7-hydrazinocoumarin

Sodium nitrite is added in small portions into a kettle containing concentrated sulphuric acid and dissolved (with cooling, if necessary), taking care that nitrogen oxides do not develop. The solution is cooled and treated slowly with a mixture of 7-amino-3-phenylcoumarin and sodium chloride. After 1 h the contents of the kettle are transferred into a vat containing a water-ice-mixture, and the mixture is stirred for 2 h. The diazo compound separates in the vat in the form of a suspension. To this is added dilute NaOH solution from a storage vessel, followed by ice, taking care that excessive local pH increases do not take place. The neutralized diazo compound is transferred with stirring into a vat containing an alkaline solution of sodium dithionite. Addition of ice is made during the introduction of the diazo compound to keep the temperature at 20 °C. After stirring, the reaction mixture is heated until all the material is dissolved, charcoal is added and the suspension is filtered through a lens filter into the isolation kettle. Concentrated sulphuric acid is then added slowly and carefully into the kettle, so that foaming over is prevented. After acidification, the contents of the kettle are kept in solution for 3 h with heating. The suspension formed on cooling is filtered through a filter press. The paste of hydrazine sulphate obtained is put into a kettle and stirred with cold water and ice, and Na_2CO_3 is added until the reaction becomes alkaline to Brillant yellow. The suspension is again filtered on a filter press, and after suction and blowing through it is transferred into casks coated with rubber.

Condensation

A paste of 7-hydrazino-3-phenylcoumarin is added to a duplicator containing acetic acid, followed by the necessary amount of acetylacetone, and the reaction mixture is heated to boiling point. After cooling to 15 °C the product that crystallizes out is filtered on a suction filter. The mother liquors are retained for the recovery of acetic acid. The kettle is rinsed with ethanol, with which the cake on the suction filter is also washed. Finally the product is washed with cold water. After thorough suction the product is dried in a drying oven. The crude product is purified by crystallization from chlorobenzene.

When hydrazinophenylcoumarin is reacted with α-oximinopropio-phenone and then cyclized by dehydration, 3-phenyl-7-(3'-phenyl-4'-methyl-pyrazol-1'-yl) coumarin (83) is obtained.

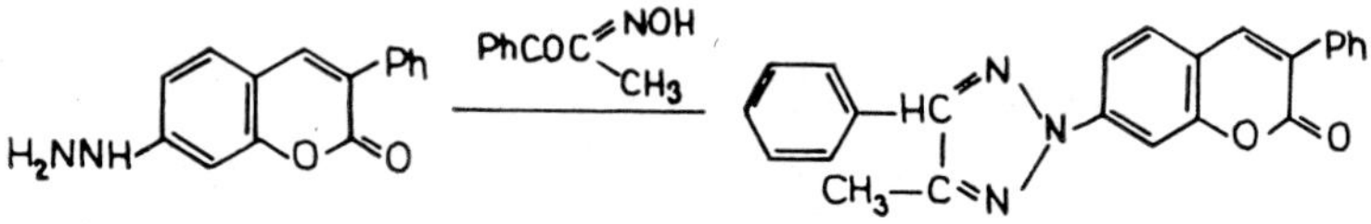

(83)

Among further reactions the preparation of 3-phenyl-7-triazolyl-coumarin (84) takes place quantitatively and the condensations of 4-methylpyrazolyl-2-hydroxybenzaldehyde with benzyl cyanide or phenylacetic acid (85) are interesting.

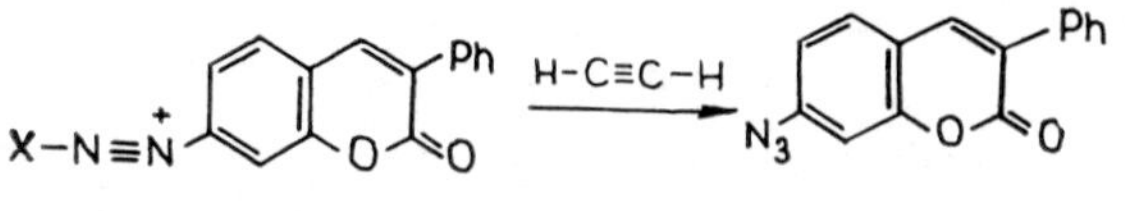

(84)

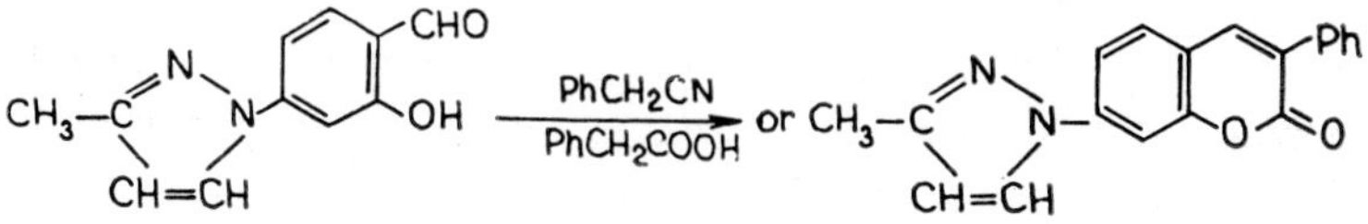

(85)

On condensation of 4-nitro-2-alkylaminotoluene with ethyl phenylglyoxylate in the presence of piperidine and subsequent reduction, aza-analogues are formed, i.e. 3-phenyl-7-amino-N-alkyl-α-naphtho-pyridones (86), which are intensive optical brightening agents with a blue fluorescence, suitable for the brightening of wool, polyamides, cellulose acetates and polyvinyl chloride.

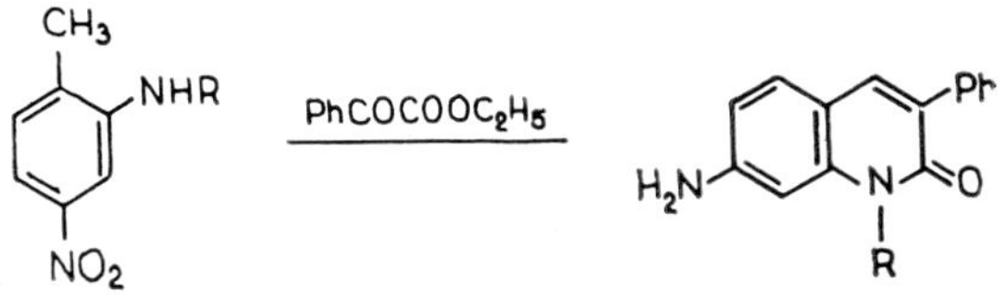

(86)

Another method of preparation consists of the reduction of 2-(N-phenylacetyl-N-alkylamino)-4-nitrotoluenes with alkali sulphides in aqueous alcohol (87).

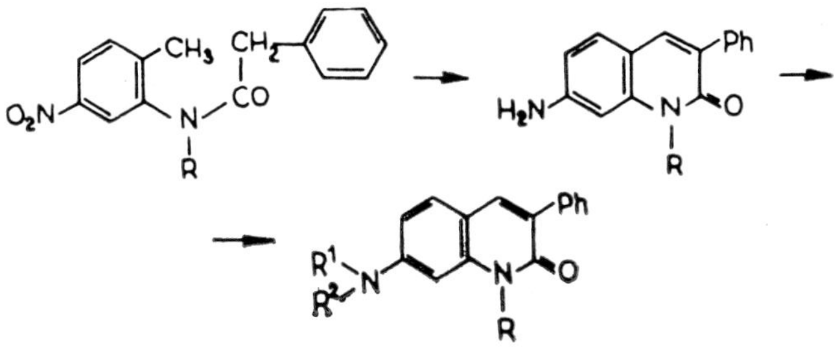

(87)

Recently derivatives of coumarone have been introduced onto the market, the optical and coloristic properties of which exceed those of coumarin derivatives. Such a substance is, for example, Hostalux HOE T 1866 (88).

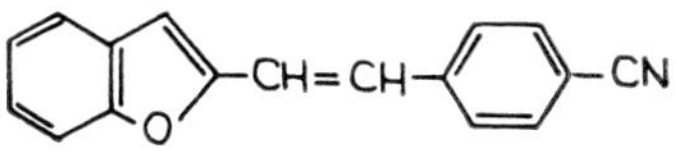

(88)

2.4   Derivatives of carboxylic acids

Numerous acids, aliphatic (fumaric acid), aromatic (benzoic, terephthalic, 1,4-naphthalenedicarboxylic and 1,8-naphthalenedicarboxylic acids), aromatic-aliphatic (cinnamic acid) and heterocyclic (2,5-thiophene-dicarboxylic and 2,5-furandicarboxylic acid), represent parent substances for the preparation of valuable optical brightening agents. They are mostly derivatives of the arylazolyl type, i.e. compounds containing a benzoxazole, imidazole and – to a smaller extent – thiazole cycle.

2.4.1 Fumaric acid derivatives

Condensation of o-phenylenediamine or o-aminophenol with fumaric acid to bis(benzimidazolyl)ethylene or bis(benzoxazolyl)ethylene gives poor yields. The condensation takes place more effectively with maleic anhydride (89).

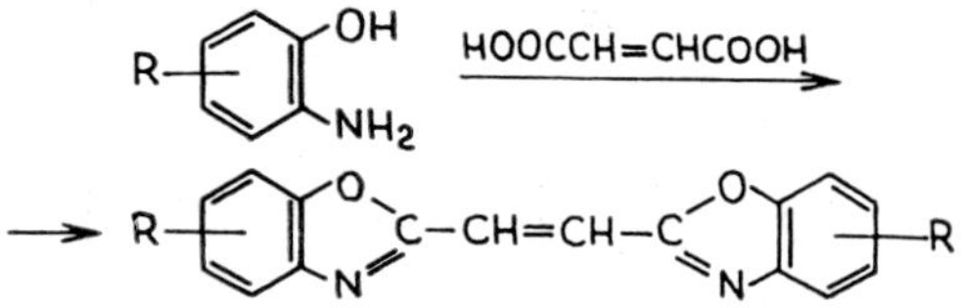

(98)

These bis-aryloxazole compounds can be prepared in good yields by condensation of o-aminophenol derivatives with derivatives of succinic acid and subsequent conversion of the product formed (90).

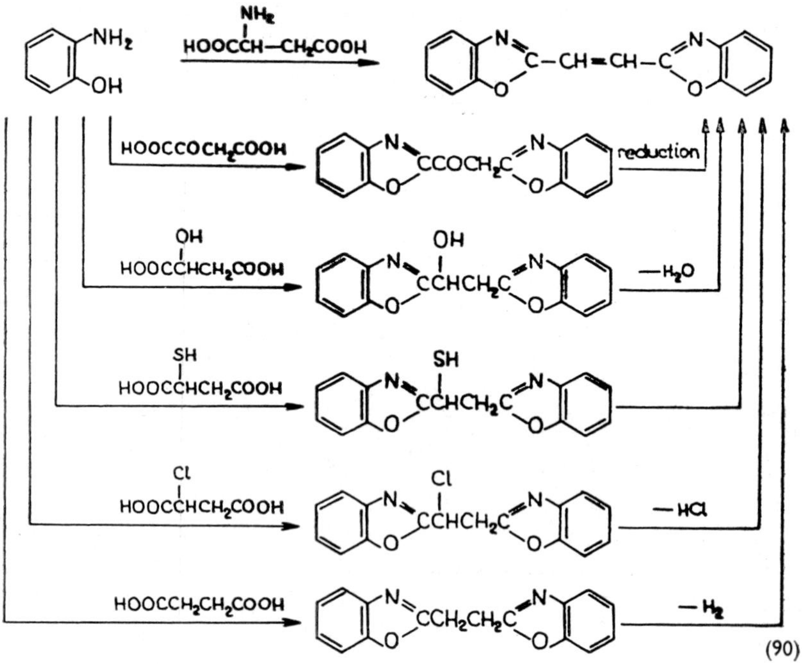

(90)

Imidazole derivatives of fumaric acid, for example Uvitex SIA (91), prepared in the same manner.

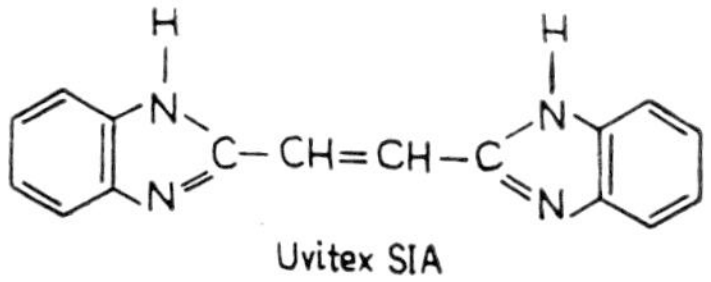

Uvitex SIA

(91)

The same preparations can be also made from 2-aminobenz-ostazole or imidazole by Meerwein synthesis (92).

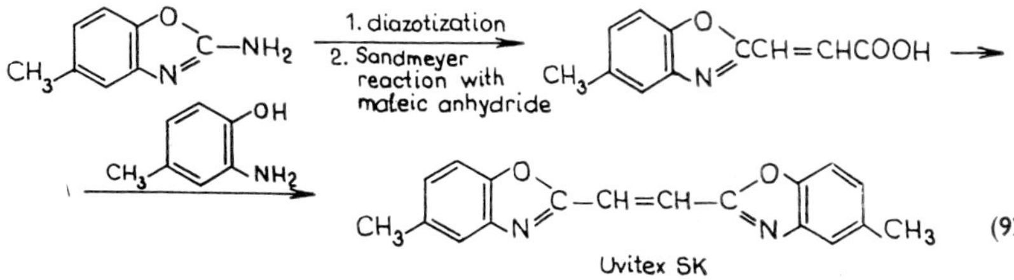

Uvitex SK

(92)

Uvitex SK is also prepared by direct condensation of 2-amino-4-methylphenol with fumaric acid (93).

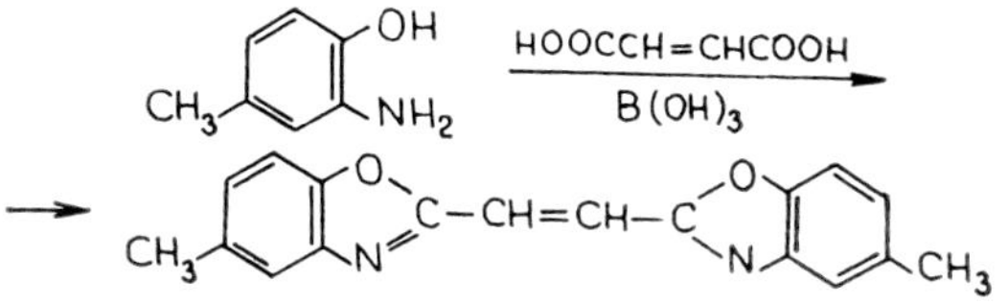

(93)

Substituents in the 5- or 6-positions of bis(benzoxazol-2-yl)ethylene shift the UV absorption from $\lambda = 355$ nm to $\lambda = 382$ nm in the following order: $-H$, $-Cl$, $-CH_3$, $-C_2H_5$, $-CH(CH_3)_2$, $-OCH_3$.

When quaternized, bis-arylimidazole and bis-aryloxazole compounds afford water-soluble optical brighteners suitable for the brightening of polyacrylonitrile. Bis-arylimidazolyl compounds can be quaternized in the usual manner using dimethyl sulphate or methyl p-toluene-sulphonate. Bis-aryloxazole compounds cannot be quaternized in this manner and a more complex procedure must be used, as shown in scheme (94).

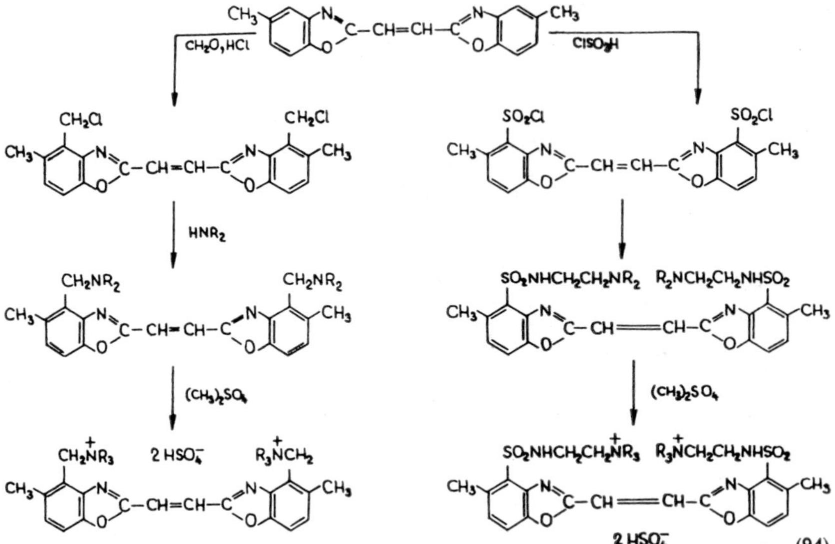

74

(94)

Technically the most important product of this type is the bis-pyridinum salt Uvitex ERN (95).

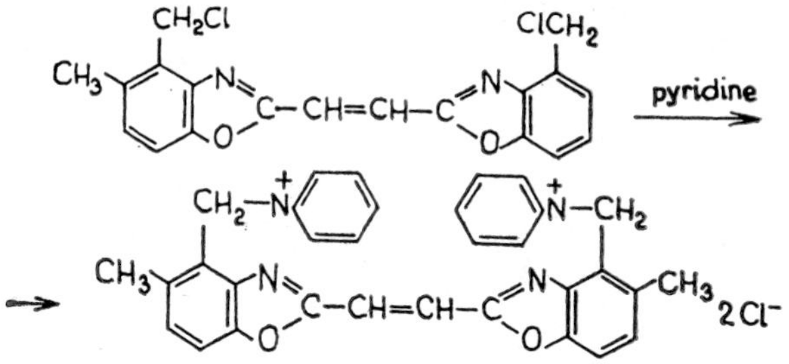

(95)

Water soluble substances of the mentioned types are also formed on introduction of one or two sulphonic groups into the molecule. Sulphonated derivatives of benzimidazole are recommended for the brightening of cotton and paper, while benzoxazole derivatives are employed for the brightening of polyamides.

2.4.2 Benzoic acid derivatives

Unsubstituted 2-phenylbenzoxazole and 2-phenylbenzimidazole are not suitable for optical whitening (thiazole derivatives are of no practical value). 2-Phenylbenzoxazole (96), absorbing at 300 to 310 nm, has been used as a UV-absorber in this region. Various acylated derivatives of 2-*p*-aminophenylbenzoxazole (acylation with acetic anhydride, benzoyl chloride, cyanuric chloride) have no practical value either. Only when aryltriazolyl, benzoxazolyl and stilbenyl residues are introduced into the molecule, are useful optical whitening agents obtained. Thus, condensation of terephthalic acid or its dichloride with derivatives of *o*-aminophenol gives rise to 1,4-bis(2'-benzoxazolyl)benzenes (97).

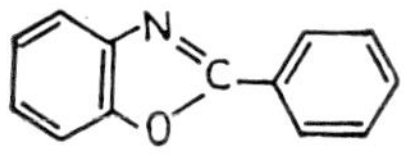

(96)

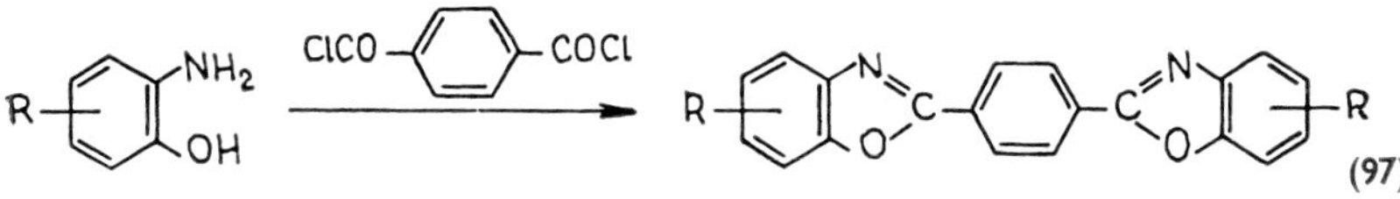

(97)

A more complex derivative of terephthalic acid, 1,4-bis(4',5'-di-phenyl-2-oxazolyl)-2-chlorobenzene, with a green fluorescence, is prepared according to scheme (98).

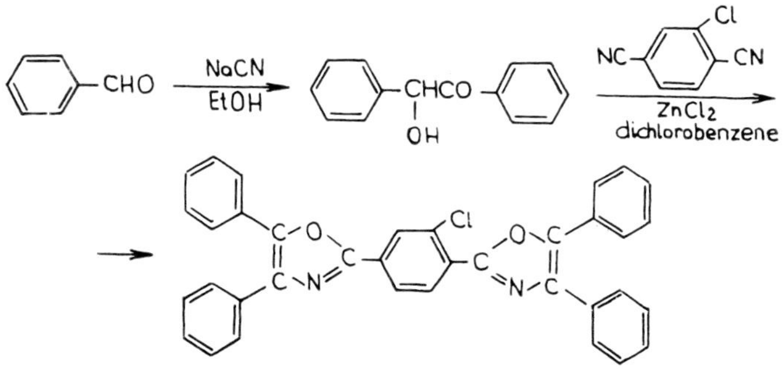

(98)

By melting *para*-toluic acid with sulphur or heating the azine of methyl 4-formylbenzoate, dimethyl 4,4'-stilbenedicarboxylate is formed. Condensation of this with *o*-aminophenol or 4-methoxycarbonyl-2-amino-phenol yields 4,4'-bis(benzoxazol-2'-yl)stilbene or 4,4'-bis(6'-methoxy-carbonylbenzoxazol-2'-yl) stilbene — (Hostalux SE) respectively (99).

(99)

Hostalux SE

Diazotization of 2-(4'-aminophenyl)benzoxazole and coupling of the diazonium salt formed with 2-naphthylamine and subsequent cyclization gives rise to the triazole derivative (100).

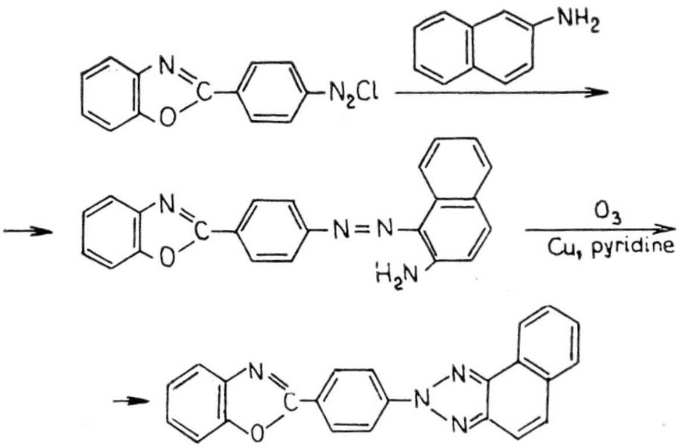

Imidazole and thiazole analogues of these compounds are prepared in a similar manner. All the compounds mentioned are used as disperse optical whitening agents for synthetic materials.

2.4.3 Cinnamic acid derivatives

This group mainly includes compounds which are derived from cinnamic acid by its simple substitution with a wide range of substituents, or more complex structures in which the cinnamic acid skeleton is an integral part of the compound, or all compounds containing the styryl residue which are not readily classified in to other groups. They can be divided into six groups: 1. cinnamic acid derivatives, 2. styrylazoles, 3. styrylbenzofurans (derivatives of coumarone), 4. styryloxadiazoles, 5. styryltriazoles, 6. styrylpolyphenyls. Of these only the derivatives of the styrylazole type are of industrial importance. They comprise a group of very wide range of compounds, the production of which is protected by dozens of patents. Compounds from the other groups have not yet achieved commercial viability.

1.   Cinnamic acid derivatives

Naphthotriazole derivatives of cinnamic acid are prepared by the coupling of diazotized 4-aminocinnamic acid with 2-naphthylamine and its derivatives, and by oxidative cyclization of the o-aminoazo dye formed to give the triazole ring (101). The carboxyl group is then esterified or converted into variously substituted amides.

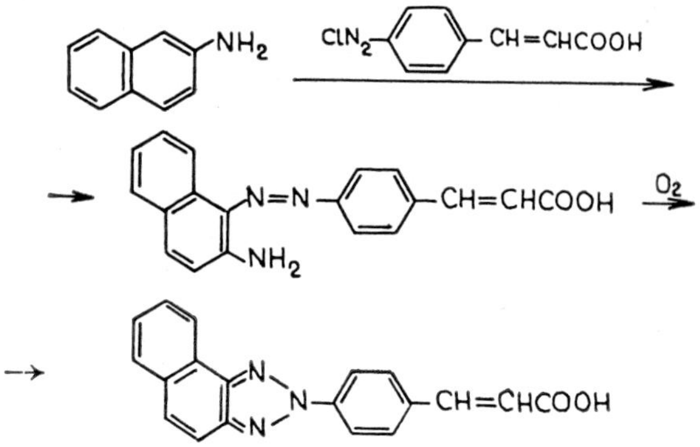

(101)

2. Styrylazoles

The group of styrylazoles can be subdivided according to the composition of the atoms in the heterocyclic substituent into styryl-benzoxazoles (102-I), styrylimidazoles (102-II) and styrylthiazoles (102-III). All three types of compounds have been patented independently.

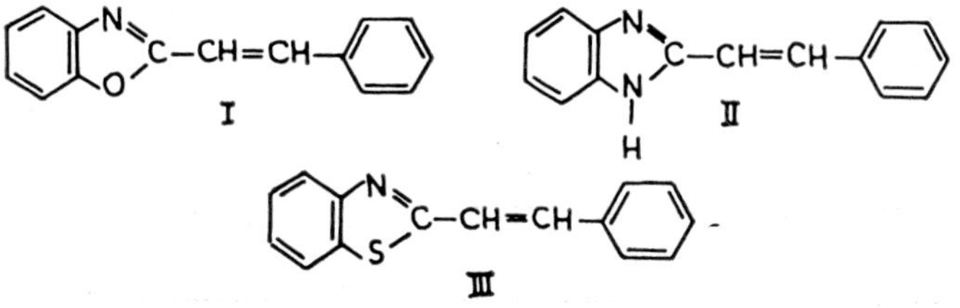

(102)

Derivatives of styrylbenzoxazole

These derivatives are prepared in various ways. One method involves the condensation of p-methoxycarbonylcinnamic acid chloride with o-aminophenol and its derivatives (103).

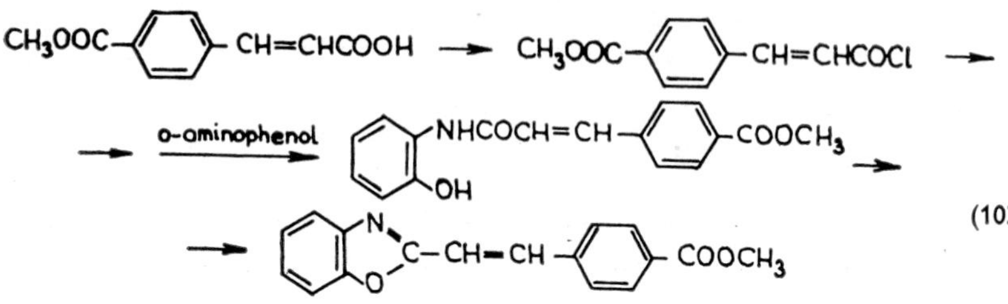

(103)

Another method is based on the condensation of aromatic aldehydes, (e.g. methyl 4-formylbenzoate) with derivatives of 2-methyl-benzoxazole (104).

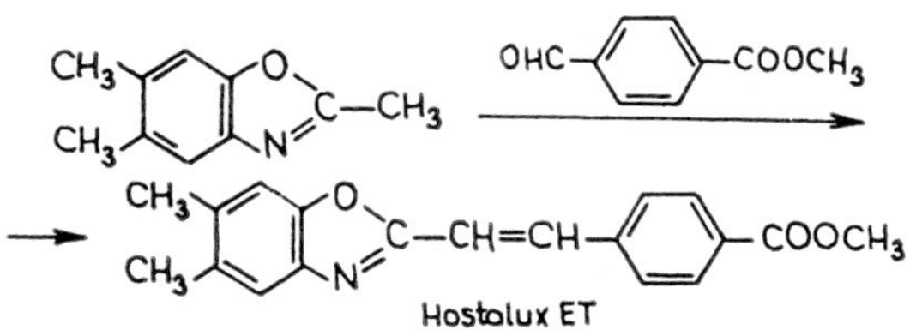

(104)

Instead of the benzoxazole derivate, corresponding acetamino-phenols can be used as starting compounds (105).

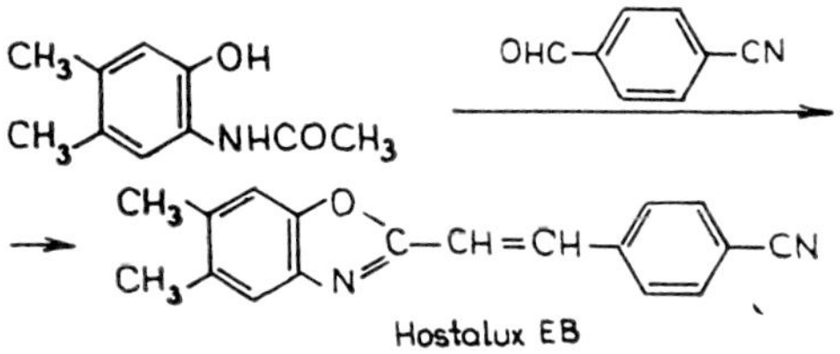

(105)

The condensation is carried out in solvents, using various de-hydrating agents if the solvent itself does not possess dehydrating properties: for example, in acetic anhydride, in piperidine, in dimethyl-formamide with potassium hydroxide, in dichlorobenzene with zinc chloride, in methanol with sodium, in xylene and dimethylformamide with toluenesulphonic acid or alkylarenesulphonates, or by melting with boric acid or with zinc chloride. Siegrist's anil synthesis takes place even with weakly reactive 2-arylbenzoxazoles, methylated in the benzene nucleus. Naturally, it is also suitable for more reactive 2-alkyl benz-oxazoles (106).

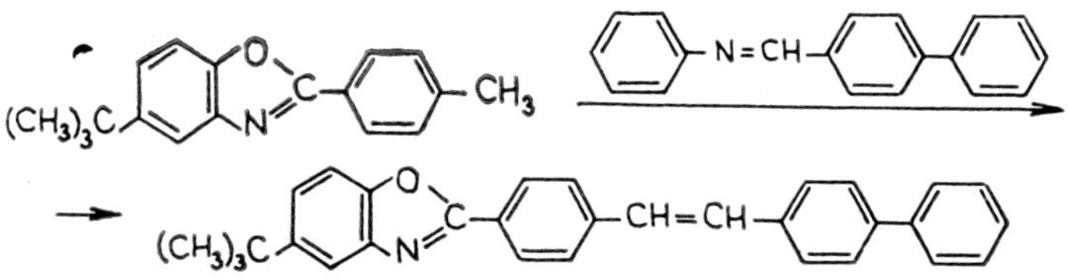

(106)

Among a great number of derivatives of variously substituted 2-styrylbenzoxazoles or 2-styrylnaphthoxazoles, Hostalux ET, Mikephor ETN, Uvitex EFT and Hostalux EB are most widely used in industrial practice. They belong to the major products used for the surface whitening of polyester materials. For exemplification, some methods for preparing agents of the Hostalux ET and Uvitex EFT type are now described.

5,6-Dimethyl-2(4′-methoxycarbonylstyryl)benzoxazole (a compound of the Hostalux ET type)

An enamelled vessel, with pressure-steam heated jacket, is filled with o-dichlorobenzene, dodecylbenzenesulphonic acid and ammonia (20 % to 23 % ammonia) and the ester of 4-formylbenzoic acid is added to the solution formed with stirring. When the ester is dissolved, the kettle is slowly heated with simultaneous introduction of an equivalent amount of 3,4-dimethyl-6-acetamidophenol. The rate of addition is regulated so that the added component has sufficient time to melt. When all the components are melted, the kettle is closed and the temperature increased to 190 − 200 °C, with simultaneous removal of water by distillation. The heating is then interrupted and the reaction mixture allowed to cool whilst slowly stirring. After cooling, ethanol is added slowly and the suspension formed is filtered on a suction filter. The filter cake is washed thoroughly with ethanol and then suction-dried. The filtrates are set aside for recovery of the solvents. The yellow cake is then dried.

1-Diphenyl-2[4′-(6″-tert.-butylbenzoxazol-2″-yl)]phenylethylene (a compound of the Uvitex EFT type)

Equivalent amounts of 2-(4′-tolyl)-6-tert.-butylbenzoxazole and 4-phenylbenzylideneaniline are dissolved in dimethylformamide with heating. After cooling to 30 °C solid potassium hydroxide is added and the suspension heated at 60 °C. When the reaction is over and the mixture has cooled, dilute hydrochloric acid is added. The separated product is filtered off under suction and washed with ethanol.

In addition to styrylbenzoxazole compounds substituted with simple substituents, derivatives have also been described with more complex substituents both in the benzoxazole heterocycle and in the phenyl group of the styryl residue, such as benzoxazolyl-, benzimidazolyl-, benzothiazolyl-, oxadiazolyl-, triazolyl- etc.

Derivatives of 2-styrylbenzothiazole

Derivatives of 2-styrylbenzothiazole substituted with simple and more complex substituents are prepared very similarly to derivatives of 2-styrylbenzoxazole. Uvitex RS (107) is an example that has been introduced onto the market.

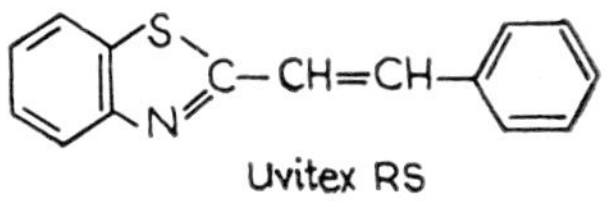

Uvitex RS

(107)

Derivatives of 2-styrylbenzimidazole

2-Styrylbenzimidazole derivatives are prepared using the same procedures as for analogous benzoxazole compounds. Derivatives of o-phenylenediamine or other aromatic o-diamines are used as starting components, rather than o-aminophenol. In some instances aromatic o-nitroamines are used and these are first condensed with the chloride of a cinnamic acid to give the corresponding amide, and the resultant nitro compound is reduced (108).

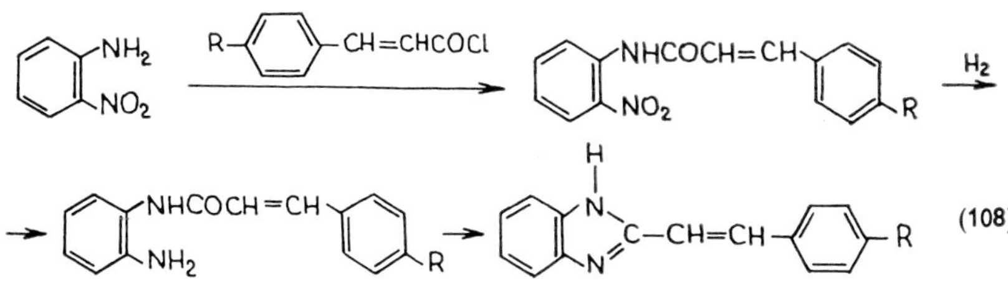

(108)

Optical brightening agents of the 2-styrylbenzimidazole type have similar applications as their benzoxazole analogues, as they are used for the brightening of polyester materials. So far no substance of this type has been used in industrial practice. Quaternized 2-styrylbenzimidazole compounds (109) have been proposed for the brightening of polyacrylonitrile materials.

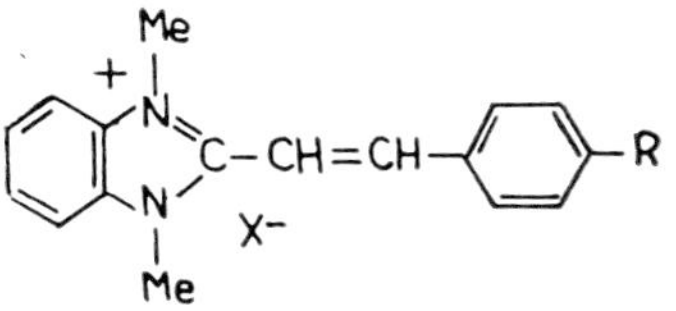

(109)

3. Styrylbenzofurans

Derivatives of 2-styrylbenzofuran are prepared by the Wittig reaction (110) from benzofurylaldehyde and its derivatives with derivatives of diethyl benzylphosphonate.

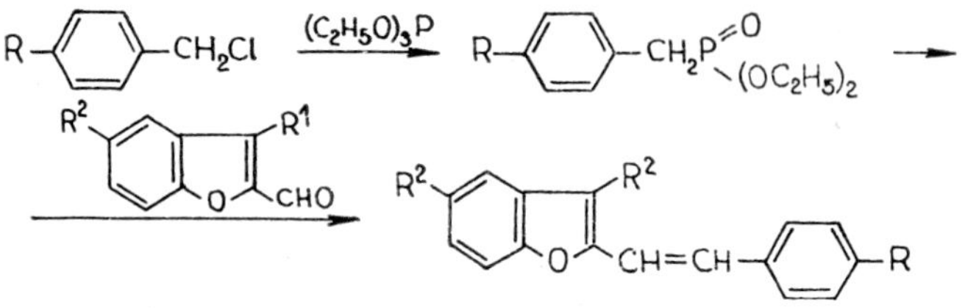

(110)

4. Styryloxadiazoles

Derivatives of styryloxadiazole (111) are synthesized by the reaction of substituted cinnamoyl chlorides with hydrazine hydrate. The hydrazide formed is condensed with benzoyl chloride and the product is cyclized with thionyl chloride.

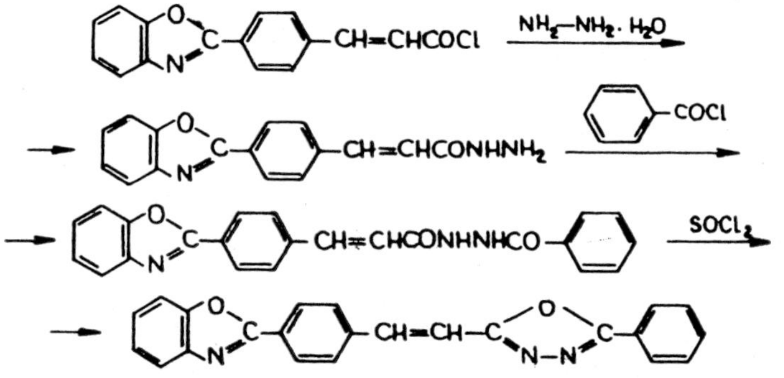

(111)

5. Styryltriazoles

Derivatives of triazole of the benzene and the naphthalene series (112) are prepared on condensation of the appropriate triazolylacetic acid with derivatives of aromatic aldehydes.

(112)

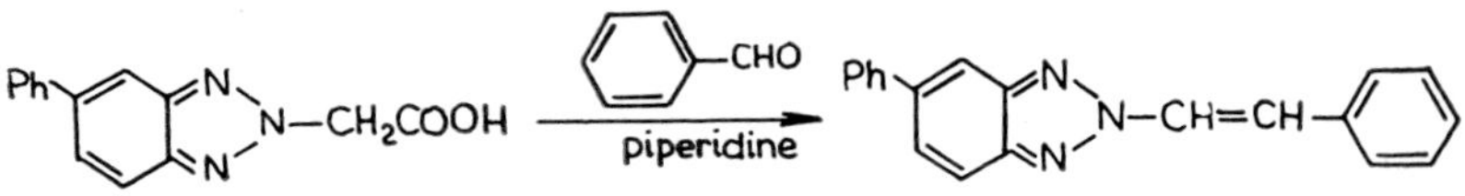

In the patent literature derivatives of styryltriazole substituted with another triazole ring (113) are described, and these are prepared by coupling diazotized aminostyryltriazole with an amine, and subsequent oxidative cyclization.

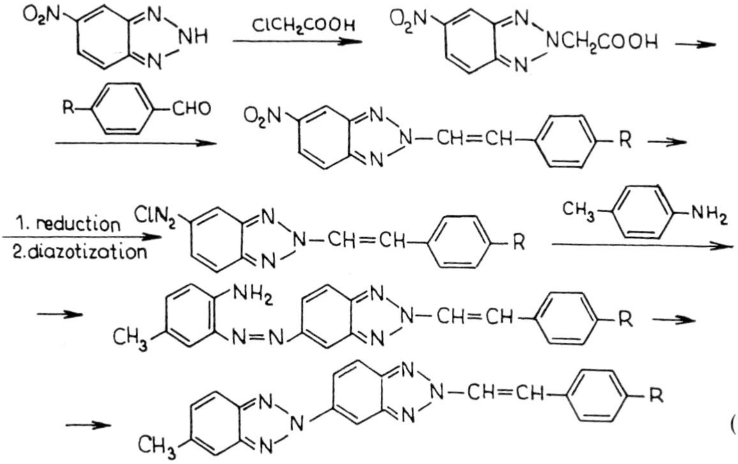

(113)

6. Styrylpolyphenyls

The phosphonate prepared from 4,4'-bis(chloromethyl)biphenyl and trimethyl phosphite can take part in the Wittig reaction with derivatives of benzaldehyde to yield 4,4'-bis(styryl)biphenyl derivatives, e.g. (114), which have been proposed for the brightening of wool, cotton, and polyamides.

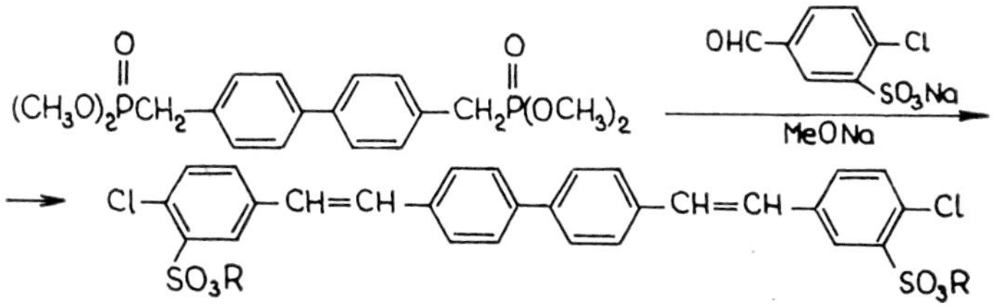

(114)

Derivatives of styrylterphenyl (115) have also been proposed as optical brightening agents.

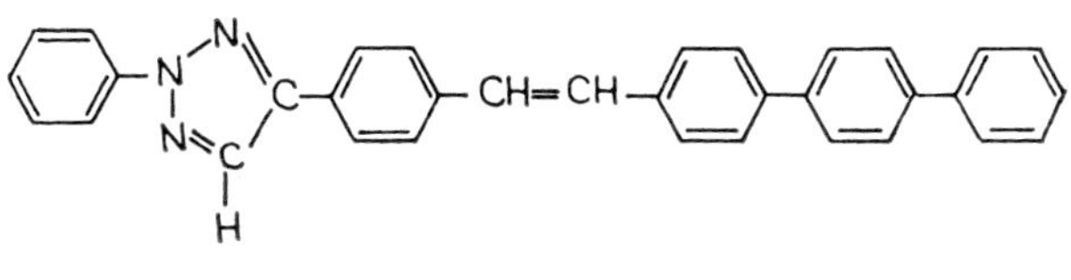

(115)

2.4.4   Derivatives of *p*-phenylene-bis-acrylic acid

Suitable derivatives of *p*-phenylene-bis-acrylic acid (116) may be prepared by condensation of *p*-phenylene-bis-acrylic acid with *o*-aminophenol. The preparative procedure starts with 2-methylbenzoxazole which is condensed with terephthalaldehyde. Compounds of this type are used for the brightening of polyesters and polyolefins, by application from aqueous dispersion or from organic solvents.

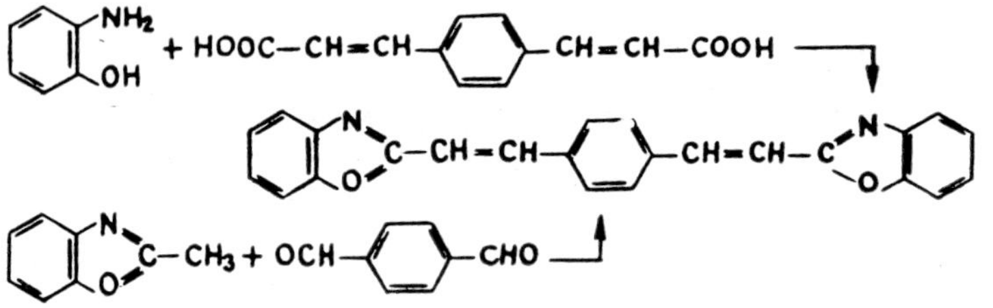

(116)

2.4.5   Derivatives of naphthalenedicarboxylic acids

Optical brightening agents based on 1,4- and 1,8-naphthalenedicarboxylic acid are of technical importance.

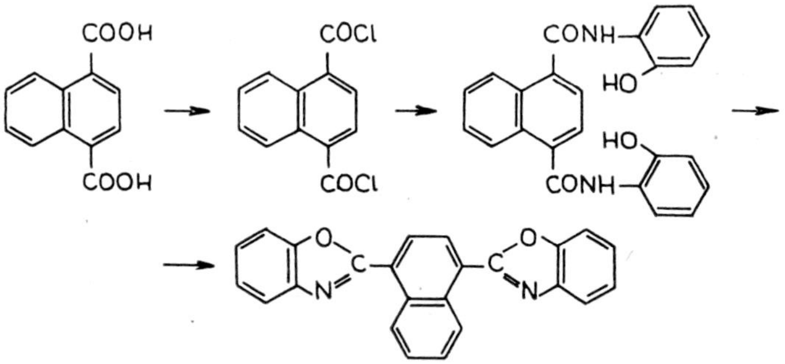

(117)

Fluorescing substances based on 1,4-naphthalenedicarboxylic acid are compounds of the oxazole type, prepared in the conventional manner by condensation of 1,4-naphthalenedicarboxylic acid dichloride with *o*-aminophenol and cyclization of the amide formed. The technically important Hoechst T 1/258 or Hostalux KCB are prepared in this manner (117).

1,4-Bis(benzoxazol-2-yl)naphthalene
(a compound of the Hostalux KCB type)

A solution of 1,4-naphthalenedicarboxylic acid in o-dichloro-benzene and thionyl chloride is heated at 100 to 110 °C until reaction is complete. Half of the chlorobenzene is distilled off under nitrogen and is replaced with fresh, dry chlorobenzene. The solution of the dichloride of naphthalenedicarboxylic acid thus formed is added rapidly to a mixture of 2-aminophenol, dimethylaniline and chlorobenzene and heated. After cooling the separated crystalline anilide is suction-dried. A product is obtained melting between 253 and 257 °C. The crude anilide is heated in trichlorobenzene with a catalytic amount of zinc chloride under nitrogen. Water is distilled off together with trichlorobenzene and the residue is cooled to 20 °C. Methanol is added, the mixture is stirred for 2 h and the product filtered off under suction, washed with methanol and dried. The bisoxazole so obtained melts between 212 and 213 °C. Compounds of this type have high melting points which, however, can be decreased substantially by alkylation in the 5-position. They serve for the dyeing of fibres (mainly polyester) in the mass. Since they display a red-violet fluorescence, they are used mainly in admixture with other optical brightening agents. Their coloristic properties are generally outstanding. The best known optical brighteners based on 1,8-naphthalenedicarboxylic acid are 4-methoxy-N-methylnaphthalimide, known under the commercial name Mikowhite AT or Mikowhite ATN, and 4-methoxy-N-butylnaphthalimide, known under the commercial name Palanilbrillantweiss FRL.

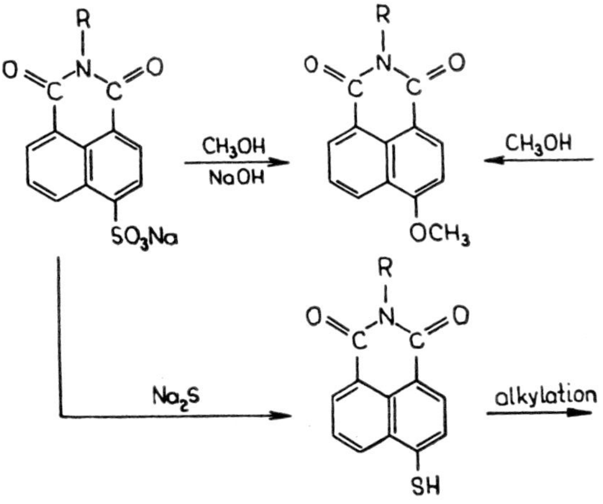

They are prepared by replacing a 4-chloro-, 4-bromo-, 4-sulpho- or 4-nitro group of an appropriately substituted *N*-methylimide of 1,8-naphthalenedicarboxylic acid with a methoxy group. The most convenient and most frequently used procedure involves heating 4-chloro- or 4-sulpho-*N*-methylimide with methanol (118a). Other procedures involve the alkoxylation of 4-substituted derivatives of 1,8-naphthalenedicarboxylic acid anhydride the alkylation of 4-hydroxynaphthalic acid, or alkylation of 4-hydroxynaphthalimide. Mikowhite AT and Palanilbrillantweiss FRL are used for whitening polyesters, cellulose acetate, polyacrylonitrile and polyolefins.

When 4-sulpho or 4-chloro derivatives of *N*-alkylphthalimide are fused with sodium sulphide and then alkylated, thio derivatives (118b) are formed. For the preparation of naphthalimide and its *N*-substituted products, naphthalic anhydride is used as starting material, which is reacted with ammonia to afford naphthalimide. The substituents in the 4-position remain intact during this reaction (119).

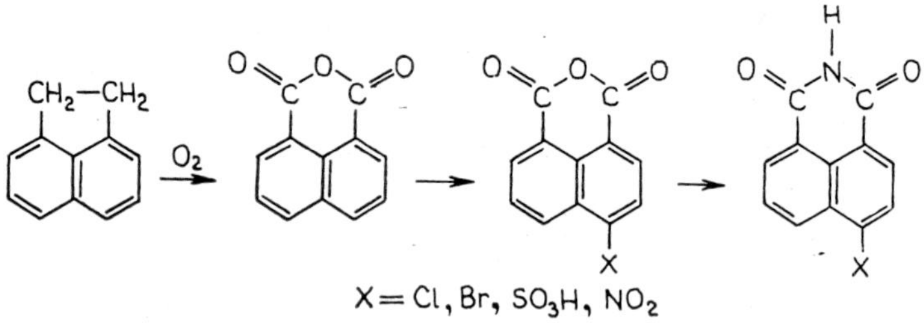

$$X = Cl, Br, SO_3H, NO_2$$

(119)

If an aliphatic amine is used instead of ammonia a *N*-substituted amide is formed (120).

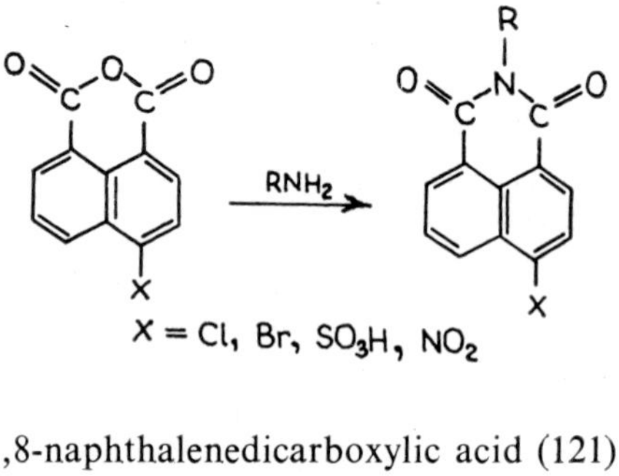

$$X = Cl, Br, SO_3H, NO_2$$

(120)

N-Alkylamide of 1,8-naphthalenedicarboxylic acid (121) are also formed on alkylation of the sodium salt of the amide:

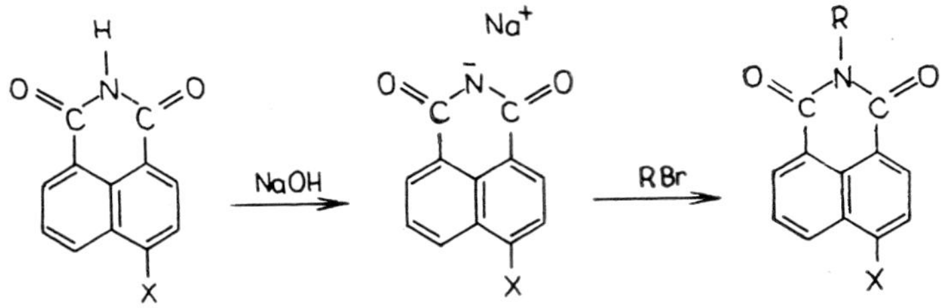

(121)

Reduction and subsequent acetylation of 4-nitro-*N*-methyl-naphthalimide gives 4-acetylamino-*N*-methylnaphthalimide, a weakly yellow product, soluble in organic solvents and fluorescing strongly. It is used as an optical brightening agent in surface-coating materials, resins, solvents, polyamides, polyesters and polyvinyl chloride. However, 4-acetylamino-*N*-butylnaphthalimide (122) is more commonly used for these purposes.

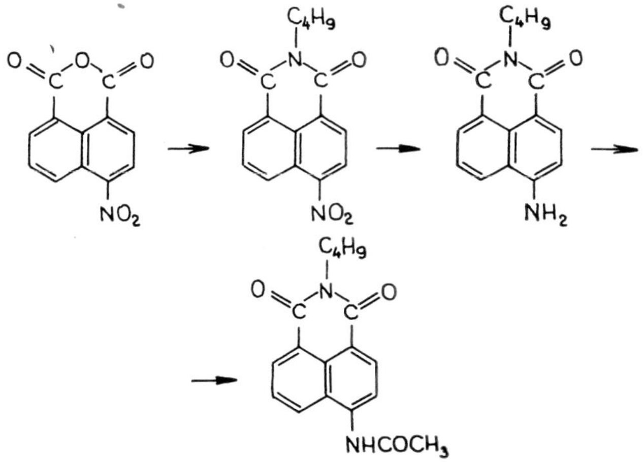

(122)

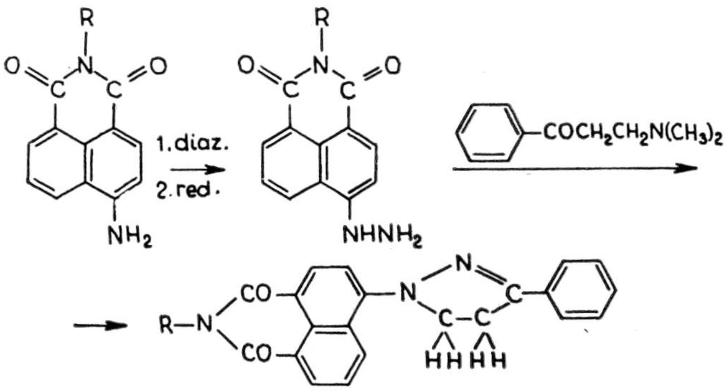

(123)

Compounds with a pyrazoline ring (123) can be prepared from derivatives of 4-aminonaphthalimide by reaction with *N*-dimethyl-2-benzoylethylamine.

2.4.6 Derivatives of heterocyclic acids

Disperse optical whitening agents of the oxazole type have frequently been developed based on monocarboxylic and dicarboxylic acids of the thiophene and furan series. Brightening agents from monocarboxylic acids are used for whitening of polyester in the presence of chemical bleaching agents, as well as for brightening polyethylene and polyvinyl chloride. The advantage of these compounds is that they may be used for the brightening of polyesters at a bath temperature up to 100 °C, so that the brightening effect can be regenerated during washing. The compounds based on dicarboxylic acids are used for the brightening of plastics — mainly of the polyester type — in the mass. Immonium salts (124), soluble in water, are suitable for polyacrylonitrile.

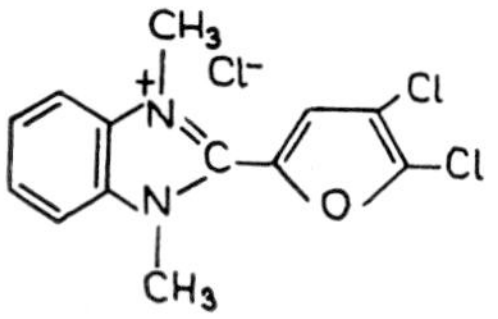

(124)

The preparation of these substances is similar to the preparation of oxazoles of other groups, i.e. an acid chloride is condensed with an *o*-aminophenol derivative to give the corresponding amide, which is then cyclized to an oxazole. The condensation and cyclization reactions can be carried out in one or in two steps. Thus, for example, condensation of the dichloride of 2,5-thiophenedicarboxylic acid with *o*-aminophenol gives Uvitex SOF (Uvitex SEB, Uvitex EBF) (125).

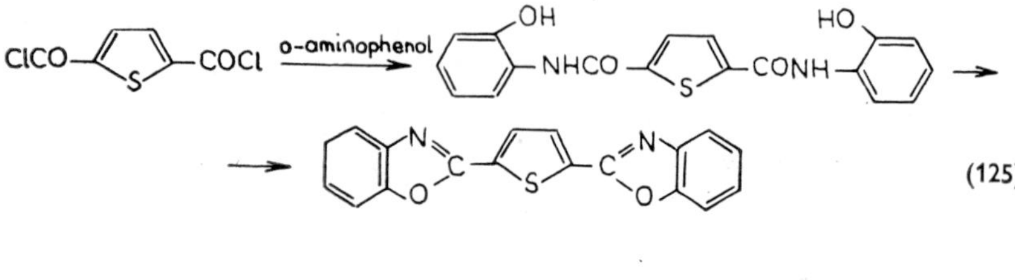

(125)

An analogous product from the dichloride of 2,5-furandicarboxylic acid is known under the name Uvitex ALN (126).

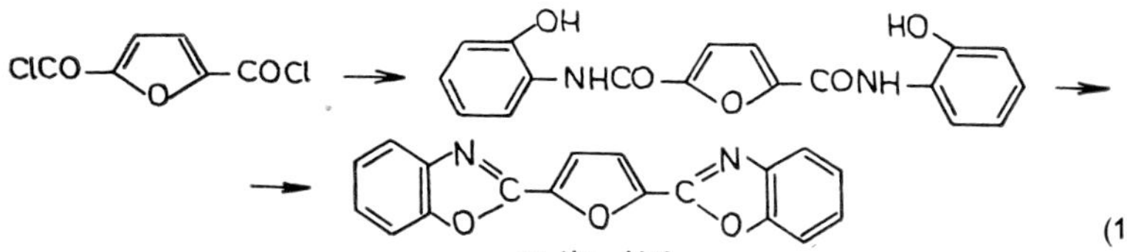

Uvitex ALN

(126)

Uvitex OB (127) is prepared from 4-tert.-butyl-2-aminophenol and the dichloride of 2,5-thiophenedicarboxylic acid, and this is the best optical whitening agent of this group and is probably (especially in the USA) the most commonly used agent of all for the mass whitening of polyester.

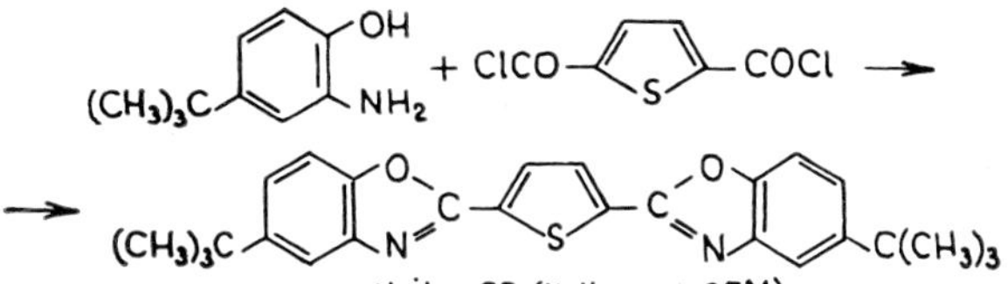

Uvitex OB (Heliophor SEM)

(127)

Uvitex 1980 (128) imparts a blue-green shade, as distinct from Uvitex OB, which imparts neutral shade.

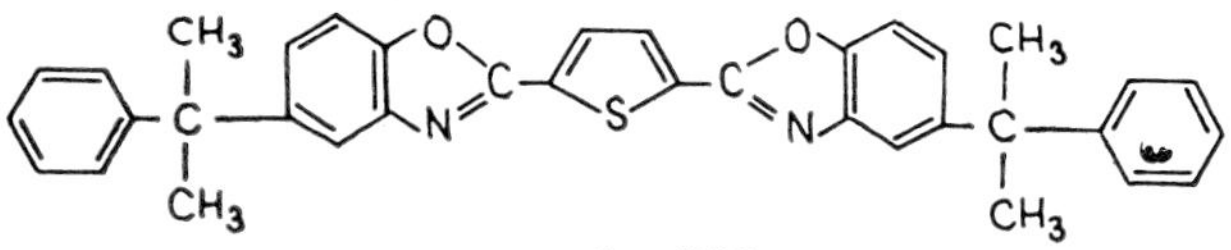

Uvitex 1980

(128)

The synthesis of these types of compounds may start from tetra-hydrothiophene-2,5-dicarboxylic acid, which is condensed with an o-aminophenol and the amide formed then oxidized [see equation (129)].

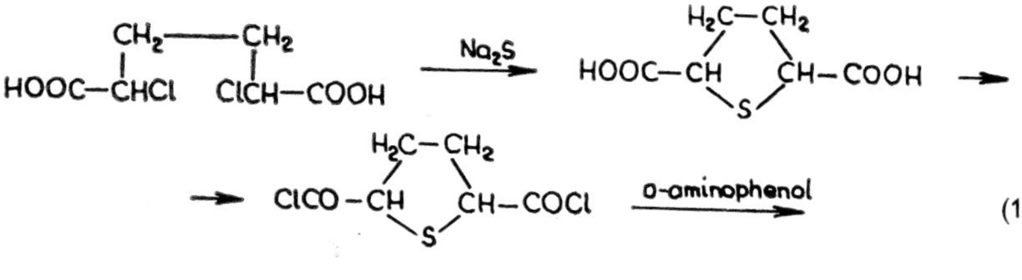

(129)

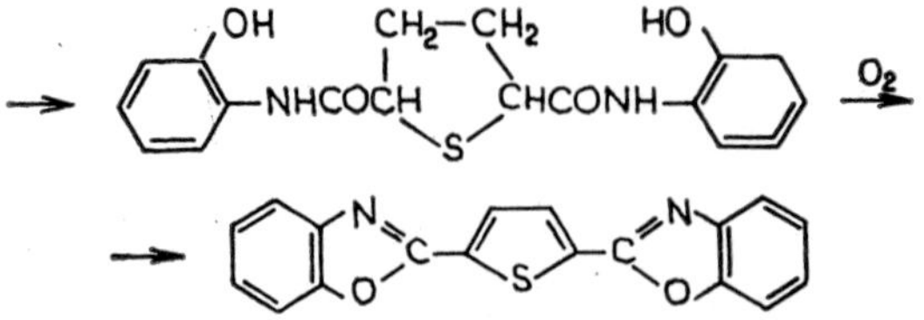

Another method of cyclization involves the dehydrogenation of derivatives of 1,4-bis(benzoxazolyl)butane with sulphur (130).

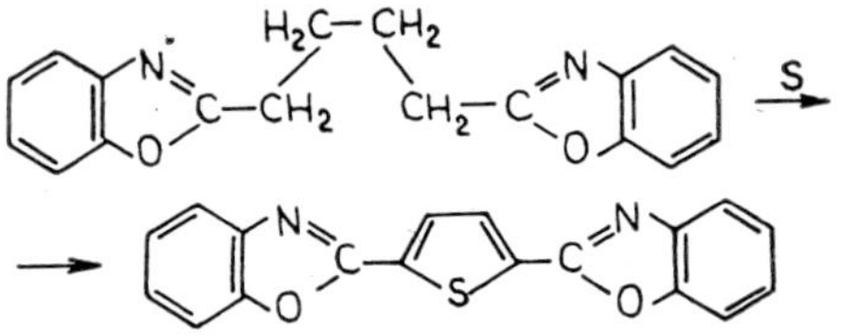

(130)

Fluorescent brightening benzoxazole derivatives of 5-aryl-2-furan- and 5-aryl-2-thiophene-monocarboxylic acid [e.g. (131)] can be prepared in a similar manner to the bis(benzoxazoles).

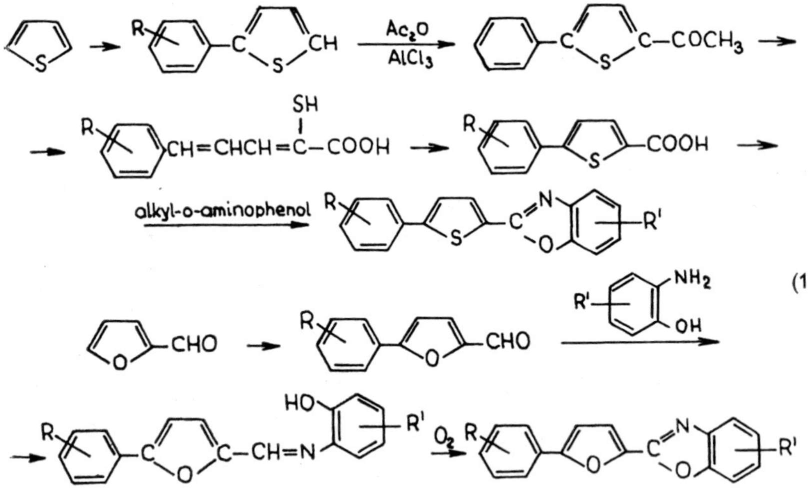

(131)

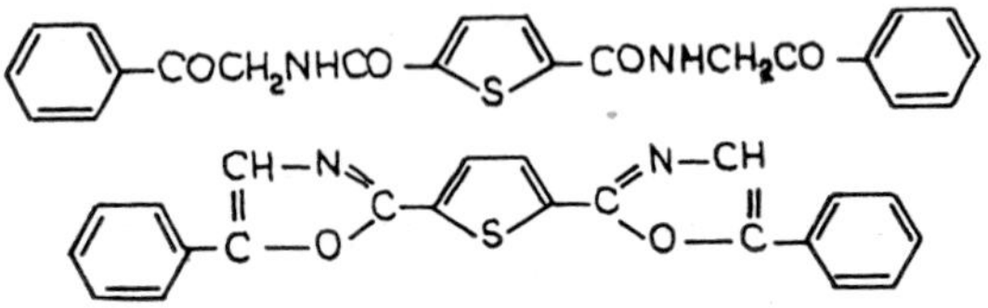

(132)

Fluorescing compounds of other types of oxazole (132) or oxadiazole (133) have also been described, and are said to be suitable for the brightening of both polyester and polyamides.

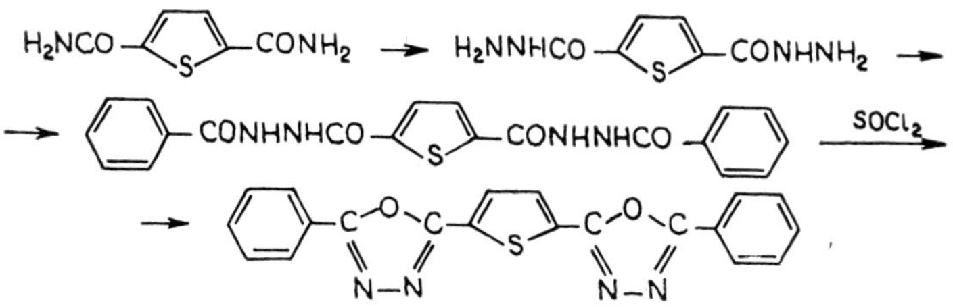

(133)

2.5 Derivatives of methinecyanines

Oxamethinecyanines (134), thiamethinecyanines (135) and their aza analogues, e.g. (136), are used as optical brightening agents.

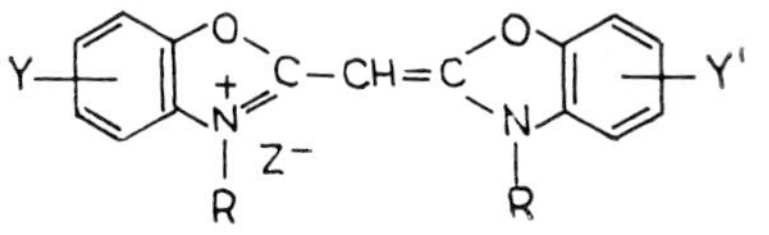

(134)

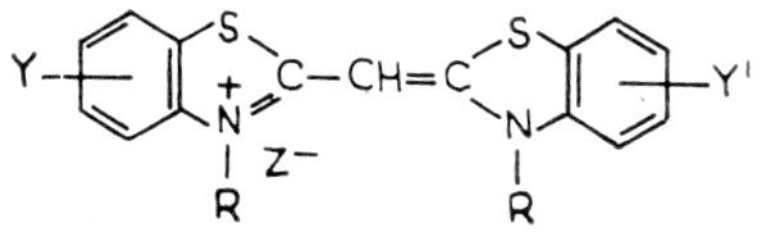

(135)

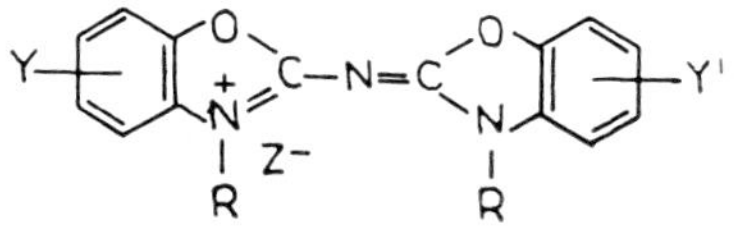

(136)

Oxamethinecyanines (137) and thiamethinecyanines are prepared by condensation of derivatives of o-aminophenol or o-aminothiophenol with diethyl malonate. In this way symmetric compounds are obtained. Asymmetric derivatives (138) may be prepared by condensation of a 2-alkylthio-3-alkylbenzoxazolium salt with a 2-methyl-3-alkylbenzoxazolium salt in the presence of some acid binding agent (sodium acetate) in alcohol or pyridine.

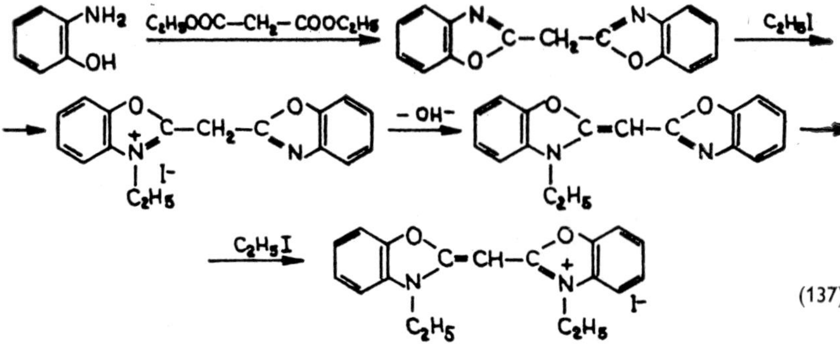

(137)

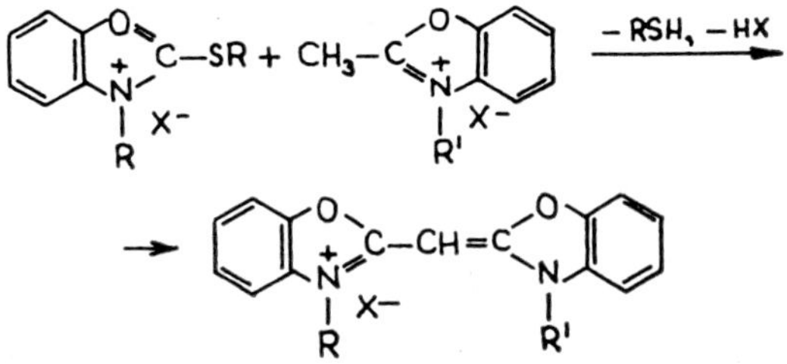

(138)

Azacyanine analogues (139) are prepared from the methyl sulphate salt of 2-methylthio-3-methylbenzothiazolium on reaction with ammonia in the presence of bases, or by self-condensation of 2-amino-3-methyl-benzothiazolium methyl sulphate in a high-boiling organic solvent. Most frequently, however, they are prepared by condensation of 3-methyl-benzothiazolimine with methyl sulphate (140).

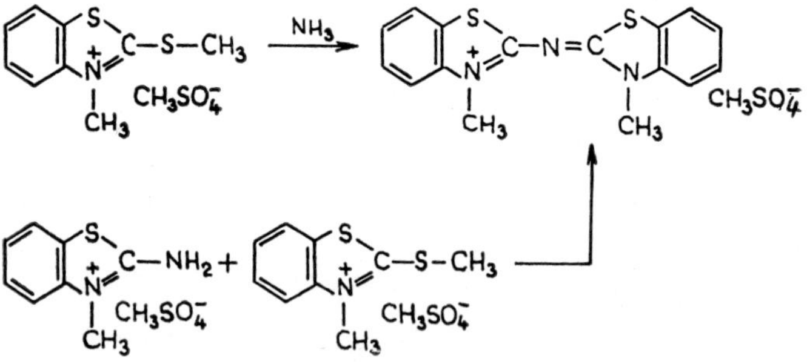

(139)

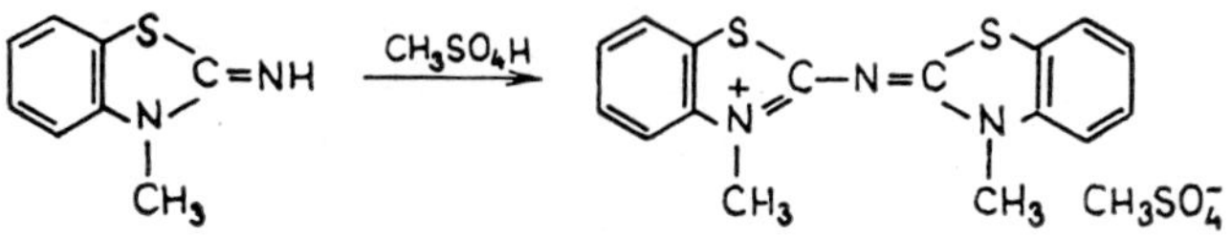

(140)

Methine compounds of this type absorb in the ultraviolet region between 370 and 373 nm and they provide suitable optical brightening agents for cellulose, polyamide and polyacrylonitrile fibres. Substitution with a methyl, alkoxy group or chlorine atom in the 5-position of the benzoxazole ring causes a bathochromic shift. Substitution of the oxygen atom by selenium causes a bathochromic shift of about 50 nm. Aza analogues of oxamethinecyanines or thiamethinecyanines show a hypsochromic shift. Oxamethinecyanine (141-I) absorbs at similar wavelengths to the corresponding thia-aza analogue (141-II).

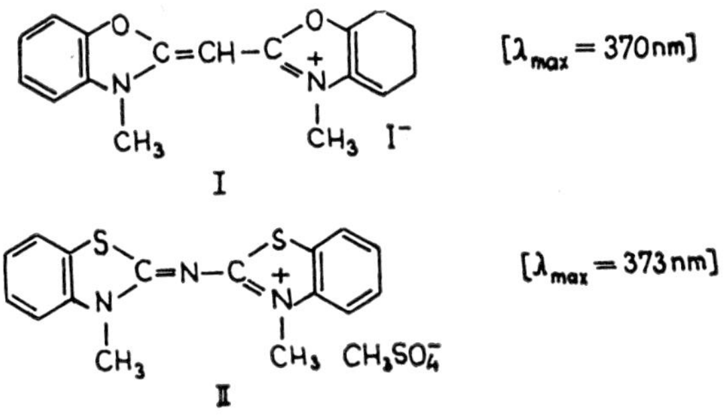

$[\lambda_{max} = 370\,nm]$

I

(141)

$[\lambda_{max} = 373\,nm]$

II

2.6 Miscellaneous optical brightener systems

In addition to compounds belonging to the previously described systems, a large number of potential brighteners have also been prepared based on anthracene, diaminopyridine, diaminopyrazine, diarylthiophene, benzidine sulphone and other structures. In the majority of cases, however they are not used in practice. However, 1-hydroxy-3,6,8-pyrene-trisulphonic acid (142), prepared from pyrene-1,3,6,8-tetrasulphonic acid by reaction with aqueous sodium hydroxide is used to a limited extent.

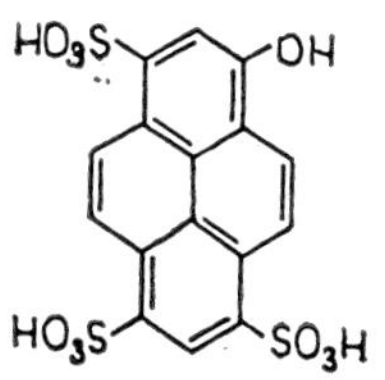

(142)

The stable, blue fluorescing 2,4-dimethoxy-1,3,5-triazin-6-yl-pyrene (Fluolite XMF) (143) is used for the brightening of polyester, and polyester/polyacrylonitrile materials. It is prepared by condensation of cyanuric chloride with pyrene and methanol in the presence of aluminium chloride:

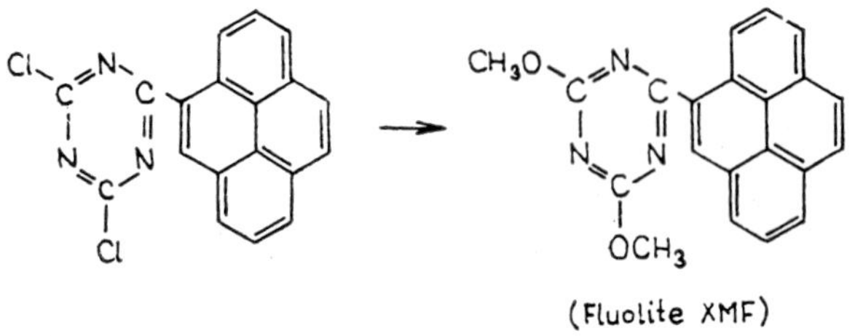

(143)

(Fluolite XMF)

2,4-Dichloro-1,3,5-triazin-6-yl-pyrene
(a compound of the Fluolite XMF type)

Equimolar amounts of pyrene and cyanuric chloride are dissolved in benzene and anhydrous aluminium chloride is added over 1 h at 20 to 25 °C with stirring. After 12 h stirring the mixture is poured into methanol, the separated product is filtered off under suction and washed with methanol. The cake is stirred with dilute hydrochloric acid at 0 to 5 °C and the suspension filtered under suction. The solid material on the filter is washed with icy water and finally with methanol. The dried cake affords a yellow product with m.p. 257 °C (from o-dichlorobenzene). 2,4-Dichloro-1,3,5-triazin-6-yl-pyrene is then refluxed with sodium methoxide solution until the starting material disappears. The mixture is cooled, precipitated, the separated product suction-dried, washed with water and methanol and dried. After crystallization from methanol the 2,4-dimethoxy-1,3,5-triazin-6-yl-pyrene obtained has m.p. 190 to 193 °C. The final product contains about 20 % of the active substance.

4,5-Diphenylimidazolonedisulphonic acid (144) is of historical interest. It was introduced onto the market under the name Blankophor WT as the first optical brightening agent for polyamide material. Today it is of no practical importance, since it has been replaced by more effective substances of a different type.

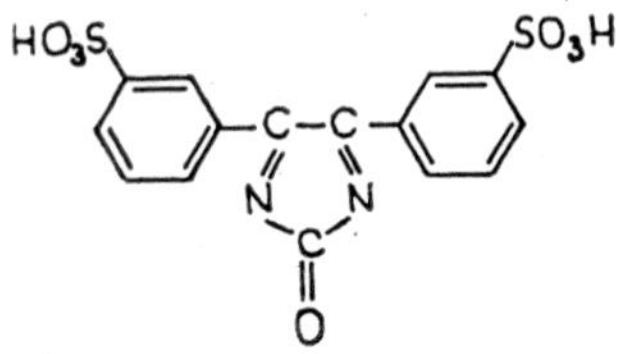

(144)

Derivatives of pyrazolinequinoline (145) are interesting optical brightening agents for polyesters and cellulose acetate.

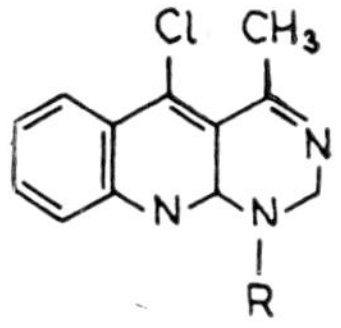

(145)

3

The Finishing of Commercial Optical Brighteners

Optical brightening agents can be used in different physical forms, depending on the substrates for which they are intended. Thus they may be obtained as powders, pastes, liquid water-soluble forms or stable dispersions. Pastes, powders and liquid forms are employed for both those brighteners containing water-solubilizing groups and those without such groups (disperse agents), whilst stabilized dispersions are suitable only for insoluble disperse optical brighteners.

3.1   Pastes and powders

Pastes are prepared by mixing the undried material, withdrawn directly from filter presses or suction-filters, with various additives, which permit the colouring power of the optical whitening agent and the physical properties of the paste to be adjusted to the desired level. The optical brightener is mixed with additives in mixers or malaxers. Sodium chloride, sodium sulphate and urea are frequently used as additives.

The powder form is produced in the same manner, with the difference that the adjusted paste is dried, after mixing, in a spray drier. The powders are produced either diluted, i.e. with considerable amount of additives, similarly to pastes, or as concentrates, where the content of additives is low or they are completely absent. The method of finishing the concentrates is determined by their intended application. They can be filled with sodium chloride, urea, dioctyl phthalate, or polyvinyl chloride. For the whitening of synthetic materials in the mass non-filled materials are used for which high purity is usually required. Quite recently so-called "instant" finishing has been used in the preparation of powders. It involves converting poorly wettable powders of optical whitening agents into easily wettable or soluble material, which is also non-powdering during addition. To achieve the "instant" finish, the wetted powder of the optical brightening agent, containing a component which on wetting becomes sticky is submitted to a mechanical treatment during which individual powder particles are brought into contact, stick together and agglomerate to fine beads, which are then dried. The resulting products are highly porous. Even a powder that is poorly wetted with water and which breaks down or dissolves with difficulty, forms "instant" finish beads which are easily wettable and which break down or dissolve readily.

The procedure for obtaining the "instant" finish is described by the following practical example.

Four parts (by weight) of the optical brightening agent (146) are added to 20 parts (by volume) of hexane and the mixture is well shaken for 5 minutes. Water (0.5 parts, by volume) is then added to the suspension and the mixture again shaken for 20 min. The beads obtained, of 0.3 to 3 mm diameter, are then separated from the two-phase system and dried. The granules can be easily sprayed, they do not dust, and they are readily wettable and soluble in cold water. If compound (147) is used as the optical brightening agent, suspended in tetrachloroethylene and then granulated with water, the same granules are obtained as in the preceding example, if the same procedure is followed.

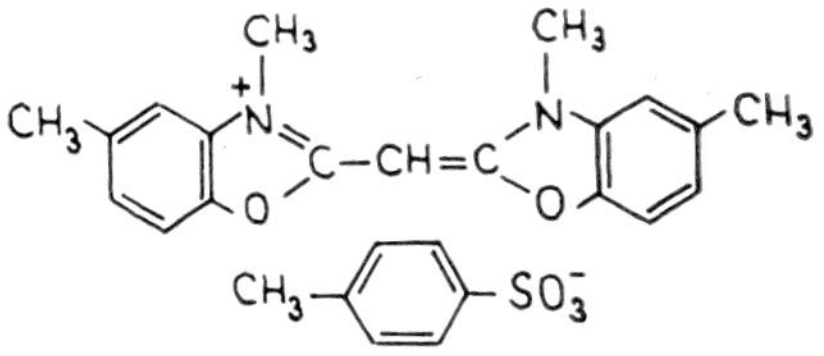

(146)

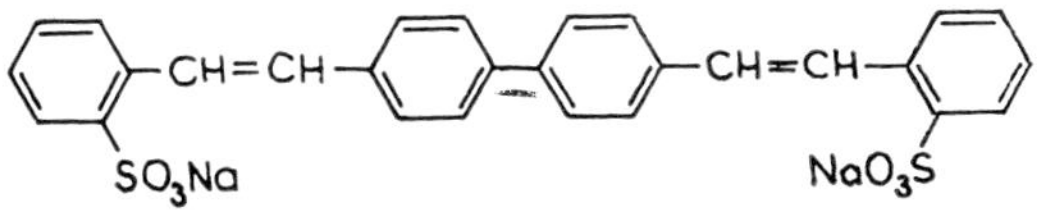

(147)

The procedures for the "instant" finish may be divided into two main groups according to the method used for wetting:

a) The substance is dried in a spray drier so that the powder still contains a small amount of water, sufficient for the "instant" finish. This is carried out directly in the spray drier and partly in a device directly connected to the spray drier. Thus the so-called semi-instant finish is obtained. The tower nozzle spray driers are used in this procedure, as here the droplets formed on spraying have larger diameters than those obtained with common spray driers.

b) The powder, usually obtained from a spray drier is humidified in the device itself by blowing in steam. The moistened powder is then agglomerated into beads with simultaneous final drying, usually in a fluidization trough. The equipment used for the "instant" finish treatment is called an *instantizer*.

3.2   Liquid forms

Liquid forms are solutions of optical brightening agents which are completely miscible with water. Both soluble and insoluble brightening agents are used in such preparations, and they are brought into solution either by purely physical processes, i.e. by dissolution in a solvent, or by chemical processes, in which the brightening agent is converted into the salt of a soluble quaternary base. Uvitex SWN provides an example of the first type, i.e. a soluble form of a whitening agent. The pure substance is dissolved in isopropyl alcohol and a dispersing agent is added to the solution so that the brightening agent is not precipitated after dilution with water. In this case a large amount of the disperser should be used. The resulting product contains about 10 % of the active substance (for example Uvitex SWN), 40 % of isopropyl alcohol and 50 % of the disperser.

The second type of liquid form is prepared most frequently from water-soluble derivatives of 4,4'-diaminostilbene-2,2'-disulphonic acid. For example, the sodium salt of the condensation product of the diaminostilbenedisulphonic acid, cyanuric chloride and the corresponding amine is converted into free acid by acidification and this is then converted with an excess of an aliphatic amine into the quaternary salt of the amine and the brightening agent (148).

The liquid forms are prepared in various ways. The most common involves the preparation of liquid forms directly from the final product. A paste or the dry product is acidified and the acid paste is suction-dried and suspended in an excess of amine. The solution is adjusted to the required concentration by addition of water or evaporation. In some instances quaternization is carried out without the isolation of the final product, or even during the synthesis, when the hydrotropic substance added serves both as the quaternizing reagent and simultaneously as a substance capable of binding the hydrogen chloride set free during the condensation involved in the reaction steps. The first modification of this procedure involves the product being precipitated from the reaction solution by salting out at 85 °C (e.g. Blankophor BA flüssig) and the oily substance obtained is isolated and mixed with ethylene glycol. Sometimes the final reaction solution is concentrated in vacuo (e.g. Tinopal 2BF),

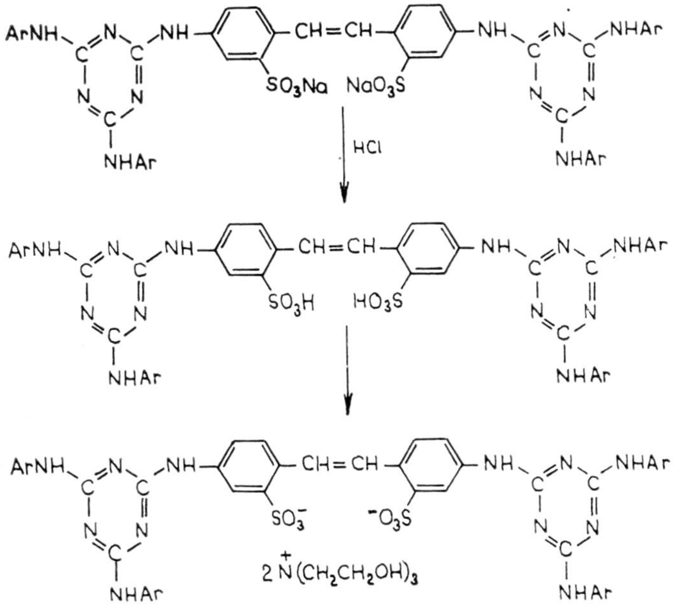

(148)

then mixed with the required amount of ethylene glycol monomethyl ether, the separated salt is then filtered off at 40 °C and diluted with water to the required concentration. In another modification of the industrial procedure the final condensation is carried out with an excess of triethanol-amine or some other organic base, and the final reaction solution is adjusted to the required concentration. As hydrotropic agents, mono-glycols, diglycols or triglycols, glycerol, various sugars, sulphite mother liquors, easily water-soluble amides, monoalkanolamines, dialkanol-amines and trialkanolamines, urea and urethanes are used. The amount of the hydrotropic substance added depends on its properties and on the properties and the form of the optical brightening agent. The liquid forms contain 10 to 40 % of the active substance and 25 up to 100 % of hydro-tropic substances, calculated on the weight of solid brightening agent.

A survey of the additives used in some optical brightening agents is presented in Table 4.

Table 4   Hydrotropic substances used in optical brightening agents

Albaphan CBS flüssig	diethylene glycol
Blankophor flüssig	urea, ammonia
Blankophor BBU flüssig	urea, ammonia
Blankophor BBH flüssig	urea, ammonia
Blankophor BE flüssig	urea, diethylene glycol
Blankophor BA flüssig	urea, ammonia
Blankophor RA flüssig	urea, ammonia, diethylene glycol
Blankophor Cl flüssig	oxyethylated phenol
Celumyl CSP flüssig	ethanolamine, diethylene glycol
Fluolite MP liquid	diethylene glycol, ethanolamine
Leukophor AC flüssig	diethylene glycol, tamol
Leukophor BS flüssig	urea, ammonia
Tinopal CH 3632	diethanolamine
Tinopal UP liquid	ethanolamine, ethylene glycol
Tinopal 2BF flüssig	diethanolamine
Tinopal 4BM flüssig	triethanolamine, urea
Tinopal LAT flüssig	oxyethylated phenol
Uvitex PRS flüss. konz.	triethanolamine, urea

To illustrate the procedure for the manufacture of liquid optical brightening agents preparative details for CC/DAS and one of the coumarin type brighteners are given.

Production of CC/DAS liquid optical brightener

Hydrochloric acid is added to a hot suspension of 4,4'-bis(1,3,5-triazinyl-2-phenylamino-4-methoxy-6-amino) stilbene-2,2'-disulphonic acid in water until the reaction is distinctly acid to Congo Red. After one hour's stirring the suspension is filtered on a suction-filter. The cake is washed with water until the acidity has disappeared and the paste is suction-dried. This is then introduced in portions into a solution of urea and triethanolamine with simultaneous heating. Heating is continued until all the material has passed into solution. The colouring strength is determined either on the basis of sample dyeing, or by titration with cetylpyridinium chloride. Depending on the result of the analysis the required amount of water is then added and − after one hour's stirring − the solution is distributed into the dispatching containers.

Production of the liquid form of 7-diethylamino-4-methylcoumarin

A kettle is filled with 2-propanol and polyoxyethylenenonylphenol ether, followed by 7-diethylamino-4-methylcoumarin, and the suspension is heated at 40 to 60 °C and stirred until both components are completely dissolved in the 2-propanol. Diatomaceous earth is added to the solution, which is filtered and the filtrate transferred into appropriate packages.

3.3   Stable dispersions

For the surface dyeing of synthetic fibres with disperse optical whitening agents aqueous dispersions are used. The active brightening substance is ground in water to a suspension with particle size 5 to 200 nm and then mixed with a suitable, mostly non-ionogenic dispersing agent. The content of the effective brightening substance in the dispersion is about 15 % and the content of the dispersing agent about 50 %.

The aqueous dispersion of the optical brightener is in some cases weakly tinted with a blue dye. The additional weak blue coloration of the dispersion brings about an increase in the brightening effect up to a maximum of 30 %. Recently stable dispersions have been prepared using complexing agents. The water-insoluble optical brightening agent is converted by means of a low-molecular complexing reagent into a water-insoluble complex which with aqueous solutions of some polymeric compounds gives suitable dispersions directly, without grinding. For example, a 1 : 1 mixture of 3-(p-chlorophenyl)-1-(p-sulphamidophenyl)-2-pyrazoline and dimethylformamide is mixed with an aqueous solution of polyvinylpyrrolidone, sometimes adding urea as well. A stable dispersion is obtained containing 35 % of the active substance, which can be diluted with water to the required concentration.

It has been claimed that practically all types of disperse optical brightening agents are suitable for this method of preparation of stable dispersions. Amides (methylformamide, dimethylformamide, dimethyl-acetamide, pyrrolidone), imides (maleimide and its N-alkyl derivatives succinimide, phthalimide), urea and its derivatives, dimethyl sulphoxide and organic acids (maleic acid, phthalic acid, tetrachlorophthalic acid, acrylic acid) are used as complexing agents. Among polymers, the co-polymers of vinylpyrrolidone, vinylpiperidone, vinyl alcohol, vinylacet-amide, vinylformamide and others are employed. The copolymers of maleic anhydride and styrene are especially suitable. For a better dispersion anion-active, cation-active or amphoteric tensides are added.

4.

Applications of Optical Brighteners

Optical brightening agents are used for the whitening of natural fibres (cotton, wool and natural silk), synthetic fibres derived from natural polymers (viscose silk, cellulose acetate) and fibres made from wholly synthetic polymers (polyamides, polyesters, polypropylene, polyacrylonitrile and polyurethane). They are also used for the whitening of paper, plastics, detergents and many other substrates. Similarly to "coloured dyes", fluorescent brightening agents can also be classified according to their dyeing behaviour as direct (or substantive), disperse and cationic groups. As far as the production volume, consumption and the number of the marketed brands are concerned, the direct brightening agents prevail. They are mostly derivatives of 4,4'-diaminostilbene-2,2'-disulphonic acid and are used mainly for the brightening of cotton, but also for the brightening of paper, viscose, linen and polyamides. Among commercial preparations Blankophor REU, BNU Tinopal BST and related brighteners belong to this group.

Acid optical brightening agents contain free sulpho groups and serve mainly for the brightening of silk and wool.

Basic optical brighteners contain amino groups, and include mainly the coumarin and pyrazoline types. They are used primarily for the brightening of polyamide (especially the pyrazoline compounds) and for the brightening of silk and wool. Among commercial preparations Uvitex WGS, Blankophor DBS 80, FBO, DCB, DCR, Belyi 2K, Delft Weiss LS, Leukophor WS, DC, Uvitex SWN, SWR and others belong to this group.

Disperse optical brightening agents are water-insoluble compounds of various structures. They are produced mainly for the brightening of polyester fibres, cellulose acetate and polyacrylonitrile. Some types of disperse optical brightening agents are used as pigments for the brightening of synthetic materials in the mass. For surface brightening of polyester the preparations Hostalux ETN, EBU, Palanilbrillantweiss R., Leukophor

EFR, Uvitex EFT, Tinopal CH 3511, ERT, ET and others are used. For spunbrightened fibres and for brightening plastics, Tinopal CH 3511, CH 3513, Uvitex MES, Hostalux SE, Palanilbrillantweiss R, Uvitex OB and Hostalux KSB are used, for example.

Cation-active optical brightening agents are compounds mainly of the methanecyanine type, which are used mainly for the brightening of polyacrylonitrile.

In addition to the direct brightening of materials, indirect brightening is also carried out by addition of brighteners to laundry powders or by some other combination with various detergents, softeners, whitening agents of inorganic origin, and other treatment agents.

From the point of view of the technological finishing of textiles, optical brighteners are colourless fluorescent dyes and their application in the textile industry and other industrial branches has become a matter of course. Together with the introduction of new types of fibres and new finishing procedures additional requirements have been placed on optical brighteners, as in the case of dyes. Optical brighteners are applied in the same ways as dyes, even though they are treated usually as auxiliary textile agents i.e. they are combined with other finishing treatments. In this respect the diffusion of the brightening agent into the substrate is an important property. The optical brightening agent must be fixed as far as possible in the outer zones of the substrate, because a maximum brightening effect can only be achieved in this manner. If the brightening agent diffuses into the interior of the molecule, the degree of brightening is decreased. The distribution of the optical brightening agent on the surface of the fibre is also important. When the distribution is unsatisfactory because the brightening agent occurs in excessively large agglomerates, a so-called greening may take place, and hence a maximum whitening effect cannot be achieved.

From the application point of view optical brightening agents can be classified approximately into eight types.

Type I. This group contains those optical brighteners that have great affinity and which are not affected by temperature and/or the presence of electrolytes. They are water-soluble and have an intense brightening effect on natural cellulose materials, paper, viscose, also polyamide and animal products. They are used in the textile and paper industry and as additives in washing powders. Chemically they include substances of the CC/DAS type, exemplified by Uvitex CF, SCF, PRS, Blankophor BA, RA, RPA, Tinopal DMS, and Leukophor BS. Substances of this type are especially suitable for application by exhaustion processes, but are less suitable for continuous methods. In solution some are un-affected even at pH values lower than 4.5; they are unstable to chlorinated compounds and metal salts, however. For textile applications they are less suitable for permanent finishes and they are unsuitable for finishing processes based on the cross-linking of cellulose materials.

Type II. Brightening agents of this type possess a medium affinity for cellulose materials (cotton, paper, and viscose), and also for polyamides and animal fibres. Their rate of adsorption is little dependent on tempera-ture and considerably dependent on the presence of electrolytes in the bath. They are soluble in water. Chemically they also belong to the CC/DAS group, and the commercial preparations called Blankophor REU, BBZ, Tinopal AB, ART and others are representative. They are used in the textile industry for exhaustion and continuous methods of application. They even resist solutions of pH lower than 4.5 but they are not stable towards chlorinated compounds. They are relatively stable towards metal salts, however, and are relatively suitable for permanent finishes, but they are unsuitable for cross-linking processes.

Type III. This group includes brighteners whose affinity decreases with increasing bath temperature, and are therefore less dependent on the presence of electrolytes. These products level very well and they are suitable for continuous methods of application. Chemically they also belong to the CC/DAS group (for example Blankophor BE extra hoch-konz., Uvitex RT, Tinopal BST).

Type IV. The brightening agents of this type are applied to paper. Chemically they belong to optical brighteners of the CC/DAS type (for example Uvitex PRS, Blankophor P, PRS).

Type V. These are optical brightening agents of medium affinity, soluble or insoluble in water and having a strong effect on polyamide, animal materials, cellulose acetate and polyacrylonitrile. They are used in the textile industry and are also added to laundry agents (for example Uvitex WGS, SWN, Blankophor DBC and FB.) Chemically they are substances of the coumarin and pyrazoline type.

Type VI. Brightening agents used for application to polyester. They are disperse, water insoluble substances used for the surface brighten-ing of polyester. Type VI is represented by preparations such as Tinopal ET, ERT, Mikephor ETN, Palanilbrillantweiss R, and Uvitex ET, which according to their chemical structure belong to the group of styryl derivatives of stilbene. These and other brightening agents are very effective when applied in continuous processes. They can be fixed efficiently on the fibres at temperatures between 190 and 200 °C and they a revery stable to light. The agents for spinbrightened fibres form a subgroup of this type. They are represented by Tinopal CH 3511 and Tinopal ERT. The substances of this subgroup are particularly effective when the polyester pulp is sprayed with them or when the brightening agent is added during polycondensation. Their stability to light is high and they are most effective in plastics made from polyvinyl chloride, polyethylene, etc.

Type VII. This group includes optical brightening agents used for brightening in the mass of polyamide fibres and is represented by Uvitex MP, the chemical composition of which has not been disclosed. Substances of this type are most active when sprayed on polyamide pulp or added to caprolactam.

Type VIII. Optical brightening agents of this type are used for the brightening of polypropylene. To brighten polypropylene in the mass, substances of the Eastman Optical Brightener OB-1 type are suitable. They are also suitable for the brightening in the mass of polyamide and polyester fibres. For surface brightening of polypropylene from aqueous dispersions, brightening agents of the Kayaphor OLN type are used. Their chemical composition is also unknown.

Survey of commercial brighteners of Types I—VIII

(The letters in brackets indicate the hue of the fluorescence: v — violet, b — blue, r — red.)

Type I
pH above
4.5

Uvitex CF (v), Blankophor BA (b), RA (v), Uvitex CFC (v), PRS (v), BF (v), S2R (v), Leukophor BCF (v), Ultraphor BP (v)

Type I-1
pH below 4.5

Blankophor RPA (v), Tinopal RP (v)

Type I-2

Tinopal DMG (v), TAS (b), Uvitex SBH (b), SBRN (b), SDM (v), Leukophor DNT (v), Blankophor BBH (v), MBBH (v), Mikephor TA (b), TM (v), Optiblanc TBU (v)

Type II
unstable to
chlorine

Blankophor REU (r), BBU (v), Ultraphor RPB (r), Uvitex CK (r), Weisstöner BBK (b)

Type-II-1 stable to chlorine	Blankophor CL (v), BHC (v), Uvitex NFW (v), Hiltamine Acid Stable (v)
Type II-2 application from peroxide baths	Tinopal 2RT (r)
Type III	Tinopal BST (v), CH 3690 (v), Blankophor CA 4451 (v), BSU (v), BKL (v)
Type IV paper in the mass	Uvitex PRS (v)
Type IV-1 surface sizing	Blankophor P (v), PU (v), BUP (v), PB
Type IV-2 paper, application by coating	Blankophor PRS (v), PSL (v), BPN (v), Tinopal BOP (v)
Type V	Uvitex WGS (r), Mikephor WS (r), Leukophor WS (r), Leukophor base (r)
Type V-1 greater fastness to light than type V	Blankophor DCB (b), Tinopal WHN (b), CH 3584 (b), Hostalux PR (v), CN (v)
Type VI polyethylene surface treat- ment	Tinopal ET (b), ERT (v), Uvitex EBF (b), Mikephor ETN (v), ETR (r), Hostalux EBU (b), EB (b), T 1/285 (v), Blankophor EBL (b)
Type VI-1 polyethylene surface treatment	Uvitex EFT (v)
Type VI-2 polyethylene in the mass	Tinopal CH 3511 (r), CH 3513 (b), Leukophor EGU (v), Uvitex MES (v), OB (b)
Type VII	Uvitex MO (v)
Type VIII polypropylene in the mass	Eastman Optical Brightener OB-1

Type VIII-1   Kayaphor OLN (v)
polypropylene,
surface
treatment

4.1   Theoretical principles of the application of optical brightening agents to fibres

Optical brighteners are applied to textiles and other materials in the same way as dyes, and the same theoretical principles and practical procedures apply for both types of colorant.

The brightening of cotton is based on principles similar to its dyeing with direct dyes. The cellulose fibre is composed of amorphous and crystalline regions. As a hydrophilic substance the cotton fibre absorbs water, and in particular the amorphous regions, 1.5 to 3 nm in diameter, are capable of sorbing molecules of the dye or the optical brightener. The bond between the dye and the fibre has been explained in terms of three essentially complementary theories. According to one of these the link between the optical brightening agent (which must have a longitudinal, aromatic, conjugated and planar molecule) and the fibre involves hydrogen bonds. The theory of binding by van der Waals forces also assumes a planar molecule, permitting the closest contact with the fibre. Bonding is due to the interaction of the dipoles of the optical brightening agent (dye) and the cellulose, and the strength depends on the number of conjugated double bonds in the dye and on the position and the mobility of the electron cloud in the dye molecule. The third theory (the aggregation theory) assumes that the dye penetrates the fibre in a monomolecular form, and having expelled a part of the water in it, aggregates there and takes on a greater volume, where it can no longer so easily leave the interior of the fibre. The planar configuration of the optical brightening agent facilitates both penetration into the fibre and aggregation of the dye.

Animal fibres are formed of proteinic substances and have amphoteric properties, rendering them capable of combining with both acidic and basic substances. In wool the groups of basic nature predominate and therefore wool is whitened primarily with acid optical brightening agents containing sulphonic or carboxylic groups. The mechanism of dyeing involves salt formation, by reaction of the sulphonic or carboxylic groups of the brightening agent or the dye with the amino groups of the polypeptide chains in the wool structure. In addition to this main process other bonds (mainly hydrogen bonds) between the fibre and the brightening agent are formed during the brightening of the wool with acid optical brighteners, depending on their structure.

In contrast to wool, in silk the acid nature of the proteinic substance (fibroin) predominates, which should permit dyeing or brightening both with acid and basic preparations. However, for silk direct brightening agents are preferred. Viscose silk (rayon) is regenerated cellulose and the same principles apply for its surface brightening as for cotton. In addition to surface brightening, brightening in the mass is also carried out, using disperse optical brightening agents.

Synthetic polyamide and polyester fibres and acetate silk are brightened with disperse brighteners in the same way as they are dyed with disperse dyes. Retention of the brightening agent in the fibre is explained by the penetration of the particles of the disperse brightening agent through the channels between the fibre molecules into the dyed material, and the formation of a solid solution in the fibre. According to an alternative, more favoured, theory, disperse optical brighteners are regarded as slightly water soluble. The dissolved brightening agent diffuses into the fibre, thus disturbing the equilibrium between dissolved and dispersed dye in the dye bath. Thus more dye will pass into solution, and can diffuse again into the fibre, and so on. The disperse optical brightening agent is retained in the fibre by van der Waals forces. Some synthetic fibres, for example polyester (Terylene) are very compact and their penetration by the brightening agent can be achieved only when dyeing is carried out at temperatures above 120 °C, so that the domestic brightening of such fibres is very doubtful, for example when textiles are laundered with detergents containing optical brighteners. Adequate affinity between the brightening agent and the fibre is an important condition for the achievement of the required brightening effect. When the affinity is insufficient, the brightening agent is only deposited on the fibre surface and imparts an undesirable greenish-yellow colour. The brightener is not very stable to light and is gradually decomposed in light, thus causing further yellowing or even browning of the substrate. With increasing concentration of the brightening agent during the dyeing fluorescence increases linearly with the logarithm of the amount of the brightening agent on the substrate, until the coloration achieves a certain limiting value (maximum effect) above which the brightening effect no longer increases, and on further increasing the concentration it even decreases. This is because the high concentration of the brightening agent forms a protective optical layer (filter) on the surface of the substrate, which prevents the excitation of the molecules of the brightener in deeper layers (so-called self-quenching, concentration quenching of fluorescence, or filter effect). The final brightening effect is very dependent on the spectral nature of the substrate. The majority of textile materials show reflectance at wavelengths between 325 and 475 nm. A very strong reflectance is observed with whitened cotton, but it is weaker in viscose and wool. All synthetic fibres absorb strongly in the near ultraviolet region. Since the fluorescence produced by optical brighteners is added to the reflectance of the substrate, the maximum fluorescence effect is achieved on those substrates whose ability to absorb in the ultraviolet region is suppressed by chemical brightening. Optical brighteners lend to the substrate a definite, though weak, blue-violet to greenish colour. This means that with dyed material, the optical brightener changes the final colour of the substrate. In the case of deep tones this effect is negligible, but in the case of pastel shades it can be very distinct. Generally, it may be reckoned that pastel yellows will appear paler, pinks will appear deeper, blues will seem clearer and stronger, greens will be bluer and clearer, and light browns and greys will be bluer.

The photochemical stability of early optical brightening agents was very low, but today at least a medium fastness towards light has been achieved.

4.2 Evaluation of optical brightening agents

Optical brighteners are evaluated in the same way as dyes. Their concentration in powder and liquid form is determined by subjective (visual) comparison of the samples in daylight or under an ultraviolet lamp, by titration (for example with cetylpyridinium chloride) or spectro-photometrically, against a standard of known concentration.

The degree of brightness and the hue is estimated visually. The skilled eye of a colorist is capable of distinguishing even negligible degrees of "whiteness" or the hues of fluorescence of the brightened sample. For this purpose the so-called white scale is used as an aid, i.e. a concentration series of a known brightening agent applied to a given material. The samples of the test brightening agent, applied at various concentrations, are compared with the standards. The so-called maximum effect is determined, i.e. the concentration of the brightener which produces coloration of optimum (maximum) fluorescence, such that when this concentration is exceeded, the fluorescence decreases. For mutual comparison of various brightening agents Ciba have proposed a white scale (originally only for textile materials, but later for plastics as well), which has 12 grades, each of which is divided into 20 parts (units). The scale is based on the coloration of magnesium oxide to which a value of 100 is given. Grades 1−4 contain diminishing amounts of yellow pigment, grade 5 contains neither dye nor brightener, and grades 6−12 contain a light-fast brightener. This scale has the disadvantage of not containing a peak white, such as that attained on polyamide and polyester, but is nevertheless an evaluation scale of practical importance for industrial practice.

A difference of twenty units is quite easily distinguishable and a trained eye can reliably discern even a difference of five units. The best results are achived in the evaluation when the samples and the scale are on the same material. The degree of brightening can also be evaluated by spectral measurement of fluorescence. The results obtained by this method are not always in full agreement with the results of subjective evaluation, and therefore it is used as an auxiliary method, especially in the determination of the concentration of the brightening agent on fibres.

As in the case of dyes, in the determination of the brightness and the hue of optical whitening agents there is interest in changing to the objective trichromatic spectral method. At the present time the whole system is still being developed and up to now no internationally accepted norm for such measurements has been accepted.

The determination of the stability of optical brighteners is still problematic and neither the testing methods themselves nor the expression of the results of the evaluation have been unified. Some firms use numerical evaluation for the expression of individual degrees of stability, others express the degree of stability verbally (low, medium, good). The numerical method is most favoured and it seems that all firms will eventually adopt it. The determination of the stability of optical brighteners on the substrate involves the same procedures accepted for dyes. Stability towards light is determined by comparison of the degree of fading of the dyed sample with simultaneously faded standards of the internationally accepted gray scale. For a maximum effect on cotton for brightening agents of the stilbene type, a light fastness grade 3 is given if the change of the brightening effect is about 20 units according to the blue scale. This change is hardly discernible and usually it must be assessed under an ultraviolet lamp. The determination of the higher fastness grade cannot be made precisely and it depends on the subjective evaluation of the observer whether it is estimated as 4 or 4 to 5. For the determination of degree of stability, scales are set in which light-fastness range from 1 to 4 is shown on polyamide, and the range 5 to 7.5 on polyester.

The determination of wet fastness presents no difficulties. The same method is used as for dyes, and evaluation involves assessing the degree of staining of brightened material which is held in contrast with unbrightened material. In general such staining is less in comparison with dyes. Evaluation of the adjacent staining caused by the brightening agent under ultraviolet light would lead to overestimation of the staining because of the high detection sensitivity of the brightening agents in ultraviolet light. Therefore the evaluation is carried out in daylight only, by comparison with an auxiliary scale of staining.

When the chemical structure of the optical brightening agents is to be determined the procedure is similar as with dyes. First the pure product is isolated from the commercial preparation, its melting point is determined and elemental analysis carried out. On the basis of its IR and NMR spectrum the basic chemical structure is determined. If the substance can be chromatographed certain results may be obtained when the unknown optical brightener is compared with a set of known preparations. Since brightening agents have a wide range of possible chemical structures, a single unified systematic procedure for the determination of their structure cannot be elaborated, as, for example, in the case of azo dyes. Each case must be solved individually.

In the structure determination of optical brightening agents the procedure for water-soluble agents is different from that used for disperse agents. The majority of water-soluble optical brighteners are derivatives of 4,4'-diaminostilbene-2,2'-disulphonic acid. In determining their structure they are first submitted to acid cleavage (i.e. the substance is boiled briefly with hydrochloric acid in cellosolve). The fragments are compared chromatographically with known intermediates (diaminostilbenedisulphonic acid, aniline, ethanolamines, etc.). For the purpose of elemental or spectral analysis the commercial product is purified by dissolution in water, clarification and precipitation (with acids). Disperse optical whitening agents are usually supplied as 15 to 20 % dispersions which in addition to the substance proper also contain dispersing agents and water, or even solvent. For the structure determination of the active substances, these must be isolated from the dispersion. The isolation is carried out either by extraction of the dispersion with solvents, or by conversion of the finely dispersed brightening agent into a coarser crystalline modification, which can be filtered off under suction. According to the first procedure the dispersion is suspended in ethanol (if necessary with addition of some other water-soluble solvent, such as dimethylformamide or pyridine) and the active substance is filtered off using a fluted filter. The solid, mostly organic-solvent insoluble material is dried and introduced into the extraction cartridge and extracted in a Soxhlet extractor, usually with chloroform. After evaporation of chloroform (by distillation) the active substance is obtained which is further purified by repeated crystallization from a solvent. According to the second procedure the dispersion is dissolved in some solvent miscible with water (acetone, dimethylformamide, pyridine), and the active substance is precipitated with water. This is then filtered off under suction and it is usually reasonably pure and can be converted into an analytically pure material more easily than in the former case.

The substance purified by crystallization is then used for spectral measurements (infrared spectra, nuclear magnetic resonance) and elemental analysis (C, H, N, S, halogens). The pure preparation is chromatographed on t.e.c. plates and compared with various known types of brightening agents. At the same time the purity of the preparation is determined chromatographically. Part of the isolated preparation must be used to detect the presence of the stilbene skeleton, by acid cleavage. The fragments are identified chromatographically.

From the results of the elemental analysis and the NMR spectra the probable types of substances are determined. A reliable assignment of the structure of the substance analysed can usually be made by comparison with the data from the patent literature. Individual parts of the analysed substance, for example substituents, heterocyclic groupings, etc. can be identified by individual procedures. The final proof of the identity of the original material with the tentative structure is carried out by the synthesis of the given substance.

Quantitative determination of optical brighteners by titration can be carried out by weighing 0.15 g of a dry or 0.3 g of a liquid product and dissolving the sample in 500 ml of distilled water. 25 ml of the solution is pipetted and titrated with a solution of cetylpyridinium chloride, until a precipitate is formed (immediately before the precipitate is formed, the fluorescence disappears). The consumption of the titration agent necessary for the sample is compared with the consumption necessary for the standard.

$$p_s = \frac{m_s \cdot 100}{m_t}$$

where p_s is the content of the fluorescing substances in the sample (percent)

m_s — content of the fluorescing substances in the sample

m_t — content of the fluorescing substances in the standard

The percent concentration, c_p of the analysed sample is then calculated from the relationship

$$c_p = \frac{m_t \cdot 100}{m_s}$$

where

$$m_t = \frac{V_c}{g_v}$$

and

where V_c is the consumption of cetylpyridinium chloride in ml and

g_v — the sample mass in g.

The cetylpyridinium chloride solution necessary for titration is prepared by dissolution of 0.75 to 0.80 g of cetylpyridinium chloride in 2000 ml of water, with addition of 100 ml of ethanol and a few drops of amyl alcohol (the latter prevents foaming).

This method can be used for the determination of the majority of preparations of the type CC/DAS.

4.3   Optical brightening agents for laundry products

More than half of all optical brightening agents produced are used in detergents. Since cotton is the most common textile material, it is those brightening agents suitable for cotton that are added to washing powders, i.e. agents of the type CC/DAS, either readily soluble in water or poorly soluble, which can also brighten polyamides and protein fibres. In mixtures with the mentioned basic types, CC/DAS optical brightening agents of other classes are also used for detergents, such as coumarin derivatives (Uvitex SWN) or derivatives of pyrazoline (Blankophor FB), which have an affinity for polyamide, protein fibres, cellulose acetate and triacetate and polyacrylonitrile. Optical brightening agents for polyester are also added to detergents – though less frequently. As already noted, effective brightening of polyester materials by surface treatment requires temperatures of about 120 °C, so that under the conditions of domestic washing (below 100 °C) the degree of brightening would be minimal.

4.3.1 Choice of optical brightening agents for laundry products

The choice of optical brightening agents for detergents depends on the amount and the type of detergent and the nature of any other components present in the detergent. For example, Uvitex PRS and substances of a similar type can only be used for anion-active washing powders, whereas the activity of compounds of the Uvitex SWN type is not affected by the ionic character of the detergent. The presence of additives also influences the choice of brightening agent. Thus, for example, agents which require the presence of an electrolyte for application to cotton cannot be added to those washing powders which do not contain fillers. Brightening agents highly stable towards bleach are necessary for the dichloroisocyanate types of detergents. The agents must also have a suitable form for their introduction into detergents and should not dust. Good solubility and dispersing ability are especially required for liquid detergents. For the preparation of powders, the form of the brightener and its particle size should be such as to ensure good miscibility. Optical brightening agents should brighten not only the washed textile, but the washing powder as well. With today's high standard of laundry products even their appearance plays an important role in the evaluation of their quality. The choice of brightening agent is also affected by the temperature of the wash water, the degree of agitation, and the ratio of the material washed to the size of the tub, since the rate of dissolution of the brightening agents in the tub depends on these factors. Optical brighteners which dissolve rapidly even at low temperature are added to detergents used for cold washing. The choice of brightening agents for commercial washing is simple from the point of view of the material washed, because the individual materials can be expertly classified. The situation is more difficult for home washing, however, where expert classification cannot be expected. Therefore several types of brightening agents must be added to the laundry powders in order to achieve the required effect. The situation is becoming even more complicated because of the increasing range of synthetic fibres now replacing cotton. While the renewal of the brightening effect on cotton is simple, the regeneration of optical brightening agents on a synthetic fibre under the conditions of home washing is problematic. Fortunately the stabilities of the brightening agents applied to synthetic materials, especially polyester, are high and the brightening effect is stable, as a rule, for the whole period of their life. In contrast to this, the stabilities of brightening agents for cotton are relatively low, but their structures do permit regeneration of the brightening effect during washing.

For the washing process Blankophor BNC, CL, Uvitex CBS and HOE 1866 are used for cellulose materials and Blankophor FB and Tinopal CH 3584 for polyamides. All brands, with the exception of substances of the Uvitex SWN type, are used exclusively for anion active laundry agents. The effectiveness of Uvitex SWN is not influenced by the nature of the detergent. They are also used for laundry products containing perborates, and are well stable to hydrogen peroxide which is set free in the bath tub. Their brightening effect does not decrease even at low temperatures and therefore they are utilizable even for enzymatic laundry preparations. Optical brighteners of the type CC/DAS cannot be applied in washing tubs containing chlorine bleaching compounds, but on fibres their stability towards chlorine is good. They display a brightening effect even in rinsing baths, and they can also be used with water softeners.

4.4 Optical brighteners for cotton

In order to achieve good results on cotton, pretreatment of the brightened material is necessary, carried out by boiling and chemical bleaching. Both the boiling and chemical bleaching processes are carried out in various combinations according to the experience and the equipment of the laundry. The highest degree of brightness is achieved by a combination of boiling in an alkaline bath at atmospheric or elevated pressure and hypochlorite or peroxide bleaching. Good results are also obtained by pressure boil during which the whitening is carried out simultaneously either with hypochlorite or with peroxide. The procedure in which the pressure boil is carried out first, followed by whitening with hypochlorite is less effective. For white material boiling under pressure is more suitable than boiling at atmospheric pressure. The hypochlorite boil, which is cheapest, is most efficient at pH 12 with a boiling time of 60 min.

When cotton is brightened, the optical brightener may be added to the peroxide bleaching bath or to the neutral exhaustion bath, but it may also be added into the winch process even in the presence of various finishing agents. Tinopal 4BM has proved to be one of the best brightening agents for peroxide baths. When applied in exhaustion and winch processes an outstanding brightening effect is obtained with Uvitex CF, Tinopal ABR and BV, Leukophor A and R. In the exhaustion process the brightening agent is added in about 0.2 % concentration. The winch process is more efficient, and in this a maximum of about 4 g of the brightening agent per litre of bath water is used at 60 to 70 % expression of surplus liquor. Brightening during a finishing process usually gives poor results. Numerous optical brightening agents are sensitive to acids and therefore often even a mild shift in pH, caused by the finishing agents, can lead to a change in hue. A change in hue can be eliminated by subsequent alkaline wet treatment. A double winch process may lead to an increase in the brightening effect. A double drawing of the whitened material through the winch is more advantageous, for example, at a concentration of 1.5 g of the brightening agent in 1 litre of bath, than at a concentration of 3 g of the brightening agent in 1 litre of bath water. The process can be adjusted so that the material is first bleached in a peroxide bath or in a winch, and after drying it is brightened for a second time in a finishing bath utilising a winch process. When held up to the light strongly brightened cotton textile materials display at most a slightly observable green hue, which can be eliminated by addition of a blue dye to the brightening bath. It is interesting that by addition of the blue dye the effect of the brightening agent can be further increased, and the type of blue dye employed is not important. Direct dyes Chlorantinlicht, Solophenyl or Solar, or mixtures of dyes Cibacrontürkisblau G (Ciba) and Alizarinviolett 2RJ (Sandoz), are suitable. Dyes are also suitable which display a weak affinity for cotton, such as vat pigments or acid dyes, (e.g. a mixture of Erioanthracencyanin IR and Erioanthracenrubin B). White cotton knitted wares should be soft to the touch and also have a brilliant hue. Thus a special finish has been devised for the brightening of tricot substrates. In this procedure a rough preliminary whitening is carried out at a liquor ratio of 10 − 15 : 1, and cross-linking is achieved at 30 °C. If the goods are well cross-linked, they are bleached at 50 °C with 15 % hypochlorite and hydrogen peroxide in one bath for 50 to 60 min. After 20 minutes' bleaching 0.2 % of the brightener is added as the bath is heated to the boil. The complete brightening process requires about 1 h. The goods are washed with warm and cold water. The whole process takes 2 to 3 h and it is carried out without changing the bath water and without the need for any stabilizing or levelling agents. For the brightening of cotton a series of various optical brighteners of the type CC/DAS is used. In addition to the already mentioned Blankophor REU, agents BBU and Hostalux CN are also used for the brightening of cotton, particularly in continuous processes with unstable and stable finishing and in combination with stable finishes based on cross-linking. Blankophor CL is also used for combined procedures involving optical brightening and chemical bleaching with chlorine compounds. These preparations are applied by non-continuous and continuous processes with all types of dyeing equipment and with perfect access of the bath water to the goods. The optical brightening proper is preceded by a pre-treatment of the whitened materials both by boiling, for the elimination of the oil residues, fat and further impurities, and by chemical bleaching. A hypochlorite-peroxide or chlorate bath is recommended for bleaching. The final degree of brightening depends to a considerable extent on the thoroughness of the pretreatment of the cotton fibre. The brightening agents mentioned can also be applied in one bath with hydrogen sulphite preparations, which increase their effect. For continuous methods of brightening cellulose materials, both wet and dry goods are used, which need not be washed after brightening. Application of substances of the CC/DAS type may be combined with alkaline peroxide bleaching, washing with soap and saponates, softening and some permanent dressings (those in which the optical brightening agents are not precipitated; otherwise the operation must be done in two baths). Optical brightening agents are used for the brightening of textile materials even during printing. The cotton textile material is brightened either before printing in the conventional manner, or after printing, when the brightening agents are added into the rinsing waters or into the reducing agent for white etchings. The light fastness properties of the optical brightening agents of the type CC/DAS used for cellulose materials, are medium in daylight and in a xenotest. Other fastness properties are: good to water and washing at 40 °C, medium to good to washing at 95 °C and to heat treatment up to 150 °C, and medium to heat treatment at temperatures above 150 °C and to sanforizing.

4.5   Optical brighteners for wool

The whitening of wool is far less important in the textile industry than the brightening of cotton. Wool is usually preferred in its natural creamy hue, rather than as a pure white. The range of optical brighteners for wool is limited. The best results are achieved with Leukophor WS or Uvitex WGS. Unfortunately the fastness of these products to light is low. In contrast, however, Tinopal RP or Uvitex RT have good light fastness, but their brightening effect is relatively low. By a combination of these two types, a good brightening effect is achieved with simultaneous preservation of good light fastness. During brightening, the material is first washed and then bleached, preferably in a peroxide bath. The bleaching may be carried out with packages without circulating the bath water, or with circulation, thus shortening the bleaching time to 4 to 6 hours. After peroxide bleaching, a 1 to 3 h reductive bleaching with dithionite (hydrosulphite) preparation is carried out, during which the optical brightening agent may be added. It is usually added into the neutral reducing bath after 75 % of the reduction time. When all the dithionite has been consumed, an organic acid is added (formic or acetic acid) which permits the exhaustion of the last traces of the brightening agent. Brightening in an acid bath alone (without simultaneous reductive bleaching) does not give such good results.

4.6   Optical brighteners for acetate and triacetate fibres

In the case of acetate fibres the stability of the optical brighteners to gas fumes is important. Therefore optical brighteners are divided into two groups, those that are stable to gas fume fading (Blankophor ACF, Uvitex ERN conc.) and those that are unstable (Blankophor DCB). Since the optical brighteners that are unstable to gas fumes give much better brightening than those that are stable, the first group is used wherever gas fume fastness is not a requirement. Some triacetate fibres contain the so-called "S-Finish". This treatment, carried out after brightening if the optical brightening agent can stand the alkaline treatment, brings about an improved stability to gas fumes when unstable agents are employed. However, the degree of brightening is somewhat decreased after this treatment.

For optical brightening of acetate and triacetate materials Blankophor DCB, DCR, Tinopal ET and ERT, Palanilbrillantweiss R, Leukophor EFR and others are used, both for exhaustion and continuous methods of application.

4.7   Optical brighteners for polyamides

For the optical brightening of polyamide fibres some brighteners may be used which are intended for cellulose. They exhaust from an acid bath at temperature above 60 °C. In addition, special agents have been developed specifically for brightening polyamides. When brightening polyamides, the type of polymer involved should be taken into consideration (for example, Perlon, Nylon, Rilsan, differ in their properties). Some types are utilizable only for a certain type of polyamide (e.g., for Perlon, Blankophor BBU extra hochkonz., BBU flüss. are suitable; for Perlon and Nylon, Blankophor RPA, Tinopal RT, WG, Uvitex NB hochkonz., Leukophor PAF). Other optical brightening agents (Blankophor DCB, DCR, CL) are suitable for all three types of polyamide. Great stress is laid on the light fastness of the brightening agents on polyamides, because they are not as easily regenerated during washing as brightening agents for cotton. In addition they are degraded by light to give products which are washed out from the material and which prevent efficient exhaustion of the new brightening agent on the polymer. Consequently the polyamide material becomes yellow. A yellowing of polyamide is also observed when material brightened with optical brightening agents of the stilbene type is for some reason or other stripped with chlorate and rebrightened. Brightening of polyamides is carried out on ordinary dyeing equipment, and is often combined with the purification of the polymer if a non-ionogenic detergent is used, since anionactive detergents decrease the degree of brightening. Prebleaching with chlorate, formerly used, is being abandoned, since polyamide usually contains a light filter which detracts from the efficacy of chlorate or peroxide bleaching processes. Reductive bleaching is used instead.

The general brightening procedure begins with a pretreatment in which the material is washed on a jigger at 40 °C. Then the material passes into the second washing bath; this washing rids the fibres of the remains of preparation agents and impurities. After washing, thermofixation follows and the material is transferred into a bath where it is bleached reductively and simultaneously whitened with optical whitening agents. The reducing agents eliminate slight dirt and the weak yellowing that arises during thermofixation. If the operation is done on a jig it is recommended that the temperature be raised slowly from 40 °C in order to prevent uneven exhaustion of the brightening agent and a corresponding uneven brightening of the material. Cooling the bath has a negative influence on the degree of brightening and the coloristic utilization of the brightener. This procedure may be used with particular advantage in pressure systems, since there is a saving of optical brightening agents and other chemicals of up to 0.6 %, as well as a shortening of the time of operation. If the temperature in the apparatus is increased above 130, °C better results can be achieved. For brightening a dyeing unit should be reserved, in which coloured material is not produced.

For the brightening of polyamides in the mass, Uvitex MES and others are used. Optical brightening agents are added to the polyamide in amounts ranging from 0.05 to 3 %, either during polymerization, or during the melting of the polyamide grit.

The pretreatment of polyamide fibres is similar to that of cotton material, and chemical bleaching is carried out in a reductive bath. Owing to the high affinity of the agents for polyamide, in order to achieve the necessary levelness, the temperature must be decreased and the pH increased, or the corresponding optical brightening agent added in several portions. For the adjustment of pH, acetic acid is used. The percentage exhaustion increases with increasing temperature (from 60 to 120 °C), with the exception of the Uvitex SWN type agents which are adsorbed at 60 °C. The material to be brightened should be damp. The light fastness of Uvitex SWN to daylight is medium, while in the case of other optical brightening agents it is low to medium; with the xenotest it is also low to medium. Fastness to water is good, and to washing at 60 °C it is good to medium. Heat fastness during heat treatment at temperatures up to 150 °C is good, above 150 °C it is medium and during fixation with hot air at 190 °C it is also medium.

4.8 Optical brighteners for polyester fibres

Polyester fibres are brightened either on the surface or in the mass, using procedures similar to that used for the dyeing of polyesters with disperse dyes, or for mass pigmentation. Surface brightening is most important. For surface dyeing of polyester materials from aqueous dispersion, optical brightening agents are usually added to the bath during the chlorate bleaching. The operation can be either an exhaustion procedure or a continuous process. Among the brands that are used for surface dyeing by the exhaustion procedure the following should be mentioned: Uvitex EFT, Blankophor EBL, Leukophor EHZ, Tinopal 3511, ERT, ET, Palanilbrillantweiss R, Hostalux ET and EBU. For the continuous method Blankophor EBL, Leukophor EHT, Tinopal CH 3511, ERT, ET, Palanilbrillantweiss R, Hostalux EBU, ET are suitable. Polyester is mass-whitened with Tinopal CH 3511, CH 3513, Uvitex MES, Hostalux SE and Palanilbrillantweiss R.

4.9   Optical brighteners for polyacrylic fibres

For the brightening of polyacrylic fibres both optical brightening agents stable and unstable towards chlorate are used. The first group includes Blankophor ACF, Tinopal LAT, Ultraphor AL, Uvitex ALN *conc.*, and the second includes Blankophor DCB *ultrafein*, DCR and ANR.

Polyacrylonitrile material occurs as three types: (a) *light*, which can be whitened without previous chlorate bleaching, (b) *yellow*, which must be prebleached with chlorate, and (c) *basic*, which is unsuitable for the production of white goods. For the first type of white polyacrylonitrile, Blankophor DCB *ultrafein*, DCR and ANR are used. The use of this type of brightening agent brings about excellent brightening results with white polyacrylonitrile materials. The yellow polyacrylonitrile is brightened with optical brightening agents (either stable or unstable to chlorine) after previous bleaching. The one bath two-stage bleaching and brightening process has proved to be the best. In this procedure the material is boiled with chlorate, using oxalic acid as activator. The bleaching is continued for 30 min, the bath is cooled to 80 °C and the excess of chlorine is eliminated with sodium dithionite. In the second phase the optical brightening agent is added and the bath again heated to the boil for 30 min. Then the material is worked up in the usual manner. By this method an extraordinarily high degree of brightening is achieved.

Polyacrylonitrile materials can also be brightened using either exhaustion or continuous processes. Blankophor DCB and DCR are particularly suitable for both types of processes.

4.10   Optical brighteners for polyvinyl chloride fibres

Polyvinyl chloride fibres are usually prebleached, using chlorate bleaching exclusively, at 60 to 100 °C, depending on the type of fibre. The low melting types of PVC permit a maximum temperature of 60 °C during optical brightening, and thus carriers must be used. During brightening the optical brightener, the carrier, a dispersing agent and a cross-linking agent must be added to the bleaching bath; (the last agent helps keep the brightener in a dispersed state). It is recommended to add powerful dispersing agents and cross-linking agents so that they will prevent settling of particles of the brightener on the fibre surface causing undesirable coloration. Carriers are mostly liquid and they should be emulsified before addition to the bleaching bath. The carrier is usually emulsified together with some of the dispersing and cross-linking agents applied. The bleaching bath should be heated over 30 to 45 minutes from 20° to 40 °C and up to 55 °C and kept at this temperature for another 30 to 45 min. Then the material is washed with cold water.

The heat-resistant types of PVC are also brightened by means of carriers or without them at higher temperatures. During brightening in the presence of carriers the procedure is the same as in the case of low-temperature material, except that the temperature of the bath is kept at 65 to 75 °C, depending on the type of carrier. This method of brightening is used for high-temperature PVC wherever lump formation needs to be prevented. In cases when this is not important (for example during the working up of flakes, brushed yarn with a high titre and stabilized yarns) optical brightening of PVC without a carrier in a boiling brightening bath suffices. In this procedure a 60 °C bath is prepared, containing an optical brightener, a dispersing agent and a cross-linking agent. The bath is slowly heated to the boil, boiled for 30 to 60 min and then allowed to cool down slowly. Among the commercial brands of optical brighteners, Uvitex WGS, Blankophor DBS 80, FBO, DCB, DCR, Belyi 2K, Delft Weiss LS, Leukophor WS and DC, Uvitex SWN, SWR and various others are used.

4.11 Optical brighteners for paper

Optical brightening agents both weaken the yellowish shade of the paper, which cannot be completely eliminated by either bleaching procedures or blueing (which leads to an even greater dulling than in the case of textiles), and they serve to replace the scarcer titanium-IV oxide. They provide excellent brightening of the paper. They are added into hollanders before the paper pulp enters the cylinders, so that a perfect mixing may take place; the amount used ranges from 0.02 to 0.2 % percent based on the weight of the paper pulp. In the case of silk and crepe paper the exhaustion type of application is preferred. The paper passes through a 0.01 to 0.05 % aqueous brightener solution. In the case of glued paper the brightening effect is lower since a part of the brightening agent is retained by the glue and the fillers. In the case of waxed paper a high effect is achieved if the waxed paper pulp is brightened by mixing with the brightener in a hollander. On the other hand, the brightening is ineffective if the waxed paper is brightened by the exhaustion process. The effect of a brightening agent can also be seen in coloured papers, where there is an increased brilliance and vividness of the hues, especially in the case of pink, blue and violet shades.

Among commercial preparations Uvitex PRS, Blankophor PRS, PSL, BPN, Tinopal BOP and others are used for the brightening of paper.

4.12 Other uses of fluorescent brightening agents

In addition to the textile, detergent, and paper industries, optical brighteners are also used in various other branches of industry. For example, optical brighteners are used for the brightening of feathers, fats, gelatine, wood shavings and sawdust, for the brightening of paints, leather, furs and straw. They are also used as light filters. In laundries and dry cleaners they are used as marking inks for invisible marking, the "cleaned" mark being visible only under ultraviolet light. The fluorescent effect of optical brighteners is made use of in the production of metallic foils. Thus anodically oxidized aluminium is "dyed" with an optical brightener, to achieve a higher degree of whiteness. The photographic industry makes extensive use of optical brightening agents. In addition to the whitening of the base paper they are also used for blueing (toning) photographs and for increasing contrasts by brightening the light parts of photocopies. They are also added into developers for coloured photographs where they improve the colour composition on the one hand and slow down the fading of colour photographs on the other. In the production of cosmetics they are added to creams in order to increase their whiteness and they also serve as absorbers of ultraviolet radiation. They are added to lubricants to increase their fluorescence and into wrapping materials (paper, plastics) for fat-industry foodstuffs, where they act as filters for ultraviolet light.

5

Toxicology

The risks of industrial poisoning are considerable in the chemical industry and this includes the production of optical brightening agents. The majority of health hazards and difficulties arise from lack of awareness on the part of the employees, who often ignore the rules that have to be observed in working with all types of chemicals. It is known that fewer health problems occur in workshops where highly dangerous substances are involved, since in such cases all hygienic measures are strictly observed, whereas in places where less dangerous substances are manipulated the safety requirements are often underestimated. Optical brightening agents are not toxic as such and they may be manipulated without any special measures, as in the case of dyes. Substrates treated with optical brighteners do not cause any harm to the skin either. In contrast, however, some intermediates used for the production of optical brighteners are more or less toxic.

5.1 Survey of toxicity of some substances used in the production of optical brighteners

Benzene

This is a transparent liquid with narcotic properties. Usually its effect does not begin with euphoria. Spasms occur and death comes rapidly. The main cause of industrial poisoning with benzene is the inhalation of its vapours. Still more serious are the chronic effects of benzene. The production of red and white corpuscles is impaired and the liver and nerves are damaged. It is now classed as a cancer-suspect agent.

Toluene

Toluene is a substance with similar properties to benzene and its narcotic effect is also similar. However, it is less toxic than benzene as it is more readily removed from the body.

Xylene

All three isomers of xylene (*ortho*, *meta*, *para*) are narcotic and irritant. Xylene is 1.3 times more toxic than benzene, but it does not affect the formation of blood. Contact with the liquid may damage the retina.

Naphthalene

A colourless solid, melting at 80 °C and boiling at 218 °C. Ingestion of 2 g of naphthalene causes death. Poisoning causes pains in the stomach and vomiting. Naphthalene also produces various eczemas by direct action and by inducing oversensitivity.

Chlorobenzene

A colourless liquid with an aromatic smell, boiling point 132 °C. It is more toxic than benzene, but it damages blood formation less. Prolonged and repeated exposure may lead to kidney and liver defects. Sometimes it provokes rashes based on oversensitivity.

o-Dichlorobenzene

A colourless liquid with an aromatic smell, boiling point 179 °C. It is less toxic than chlorobenzene. At higher concentrations its vapours irritate the eyes and upper bronchial tubes. On the skin it causes blisters; it is dangerous to the eyes and it is absorbed by the skin.

p-Dichlorobenzene

A colourless substance with m.p. 54 °C. Causes slight headaches. The vapours are irritant.

Trichlorobenzene

A solid substance. It occurs either as the vicinal isomer (m.p. 54 °C), an asymmetric isomer (m.p. 17 °C) or a symmetric isomer (m.p. 63 °C), or as a mixture of all three isomers. All exhibit acute or chronic medium toxicity on ingestion or inhalation.

Chlorotoluene

A colourless liquid boiling at 160 °C. Its vapours are toxic. No data are available on the effect of either the mixture or individual isomers.

Benzyl chloride

A colourless liquid boiling at 179 °C with strongly irritant proper, ties; it may even cause lung oedema. It is toxic and absorbed by the skin-damaging the eyes and burning the skin.

Benzal chloride (α,α-dichlorotoluene)

A colourless liquid boiling at 214 °C. It has similar properties to benzyl chloride but it is less toxic and irritant

Benzotrichloride (α,α,α-trichlorotoluene)

A colourless oily liquid of b.p. 214 °C. Unpleasant-smelling, not toxic on ingestion, toxic on inhalation and extremely dangerous to the eyes and skin. It has a medium irritant effect, causing eczemas.

p-Xylylene dichloride (1,4-bis(chloromethyl)benzene)

A white solid, melting at 100 °C. Burns the skin.

Methanol

A colourless liquid with boiling point 64 °C, hardly recognizable from ethanol by the sense of smell. The narcotic effect is slightly less than in ethanol, but the feeling of intoxication lasts longer (the excretion of methanol from the body is slower). It is toxic, ingestion of 5 to 10 ml causes serious poisoning and amounts exceeding 30 ml cause death. The poisoning begins with nausea, vomiting, convulsive pains in the stomach, cyanosis and respiration difficulties. It damages the blood vessels, liver and kidneys. The most important consequence of poisoning is damage to the optic nerve, which is always bilateral, irreparable and leads to blindness. Chronic intoxication with methanol is less dangerous than acute. On inhalation the irritant effect is mainly observed, followed by head-aches, shivering, neuritis, and (on inhalation of strong concentrations) damage to the sight.

Ethanol

A colourless liquid, boiling at 78 °C. Rapidly absorbed into the body, causing intoxication. Ingestion of 300 to 500 g causes death (i.e. 6 to 8 g per kilogram of body weight). The inhalation of ethanol vapours causes irritation and intoxication. The skin is damaged as with other solvents (defatting effects). It increases the sensitivity of the organism to poisoning.

Propanol

A colourless liquid boiling at 97 °C. Little toxicity on ingestion or inhalation; dangerous to the eyes, not dangerous to the skin.

2-Propanol (isopropyl alcohol)

A colourless liquid with b.p. 82 °C. Similar to propanol. It is about twice as toxic as ethanol and also about twice as narcotic. It is absorbed by the skin and it may cause temporary disturbance of liver and kidneys.

Ethylene glycol (1,2-ethanediol)

A colourless liquid boiling at 197 °C, considerably poisonous on ingestion. It affects the central nervous system, kidneys, blood vessels, causing acidosis. Severe poisoning occurs after ingestion of about 59 ml

of ethylene glycol, the lethal dose is 100 ml. The vapours are less dangerous, weakly irritating, absorption by the skin is low.

Formaldehyde

A pungent smelling gas, occuring usually in the form of a 40 %
aqueous solution, containing also methanol and formic acid. Formaldehyde is a protoplasmatic poison and it is rapidly absorbed through
the lungs, skin and digestive tract. Acutely it causes local irritations, its
vapours irritate eyes and upper respiratory tract, but lung oedema or
spasms of the vocal cords may also occur. Formaldehyde gives warning
of acute poisoning by its penetrating smell (it can be smelt at a concentration of 0.0002 mg l^{-1}). The danger of chronic poisoning is debatable.
In contact with formaldehyde vapours or solutions the skin hardens and
becomes coarse, the nails become soft and fragile.

Benzaldehyde

A liquid boiling at 178 °C and smelling like bitter almonds. It
has irritant properties and it is easily absorbed through the skin, causing
local anaesthesia. Death occurs on ingestion of 50 to 100 g of the substance.

Cinnamaldehyde

A liquid with b.p. 252 °C, with strongly irritant properties. On
ingestion the mucous membranes and digestive tract may be attacked.

Acetone

A colourless liquid with an agreeable smell, boiling at 56 °C. On
ingestion or inhalation it is not toxic. Its vapour irritates the eyes and
respiratory tract, more extensive exposure causes headache. The symptoms
of poisoning appear only several hours after inhalation. On the skin
acetone causes slight irritation.

Formic acid

A colourless liquid with a pungent smell, boiling point 100 °C. It has a strong irritant effect, especially on the upper respiratory tract and eyes. It burns the skin and at higher concentrations (above 7 %) it causes painful burns and blisters. It also increases deterioration of teeth.

Acetic acid

A colourless liquid with a pungent smell, and boiling point 118 °C. It is only weakly toxic or non-toxic on ingestion or inhalation. It is dangerous to the skin and very dangerous to the eyes. It is strongly irritant and causes inflammation of the respiratory tract and conjunctivitis and it also increases deterioration of the teeth.

Benzoic acid

A colourless substance with m.p. 122 °C (sublimation), mildly irritant and acutely mildly toxic. Its vapour irritates the skin and respiratory tract. It also causes nausea.

Phenylacetic acid

A colourless substance with m.p. 77 °C, reported as toxic. The mere smell can induce a reaction with sensitive people.

Phthalic acid

A colourless substance with m.p. 206 °C, practically non-toxic, which may, however, cause eczemas.

Dimethylformamide

A colourless liquid with b.p. 153 °C irritant on ingestion or inhalation. In some persons it causes headaches, loss of appetite and digestive troubles. It is absorbed by the skin.

Phenol

A white substance with m.p. 42 °C (b.p. 181 °C). It possesses corrosive properties and after absorption it is toxic. When the skin is splashed with phenol burning is first felt, then the affected site becomes insensitive and the skin whitens. After short contact with the substance the skin peels and heals well. After more severe damage even gangrene may develop, which heals with great difficulty. Phenol is absorbed by the skin. Solutions of a concentration higher than 10 % cause severe burns. When it is in contact with the skin the time of contact is more important than the concentration. Acute poisoning has the following symptoms: giddiness, buzzing in the ears, suffocation, paleness, sweating, headaches. Death may occur as soon as one hour after ingestion or after severe contamination of the skin. The lethal dose is 0.15 to 1.0 g per 1 kg of body weight. For an adult the lethal dose is estimated at 8 to 12 g. Chronic effects are not serious. It is reported that phenol causes headaches, irritation and sleeplessness.

o-Cresol (3-methylphenol)

A white substance with m.p. 31 °C, with similar properties to iphenol, but less toxic. The same is true of m-cresol and p-cresol. m-Cresol is less toxic than the o-isomer and p-cresol is less toxic than m-cresol.

1-Naphthol

A white substance with m.p. 122 °C, less toxic than phenol. It irritates the skin and mucous membrane and it is absorbed by the skin. Poisoning may occur even after inhalation, manifesting itself by abdominal pains, vomiting, convulsions, fainting and inflammation of the kidneys.

2-Naphthol

A white substance with m.p. 122 °C, with the same properties and effects as its 1-isomer. It is about half as toxic as 1-naphthol.

Nitrobenzene

A yellowish liquid, b.p. 211 °C. Acute poisoning causes cyanosis, convulsions and respiratory difficulties. In cases of very severe poisoning death occurs by paralysis of respiration. The poisoning is considerably worsened by alcoholic drinks. Acute poisoning may be caused by ingestion, inhalation or absorption by the skin. The lethal dose is 1 to 5 g of nitrobenzene.

Nitrotoluene

A yellow liquid with b.p. 222 °C, having substantially the same effect as nitrobenzene. m-Nitrotoluene (b.p. 230 °C and p-nitrotoluene (m.p. 51 °C) are less toxic.

Aniline

In the pure state it is a colourless, slowly yellowing and solidifying liquid with b.p. 184 °C. It causes cyanosis and it has a direct effect on the central nervous system. On acute poisoning cyanosis appears first on the lips, then on the ears, nose and nails. In the case of slight poisoning symptoms other than cyanosis do not appear. With more severe poisoning, cyanosis is accompanied by drowsiness or even fainting. Death only occurs after severe exposure. Death may occur after ingestion of 1 g of aniline, but otherwise the lethal dose reported is 25 g. In industry combined poisonings usually take place, i.e. simultaneous inhalation and absorption through the skin. Sensitivity to poisoning is increased by heat, solar radiation and alcoholic drinks (even weak beer). Aniline causes chronic poisoning accompanied by damage to the nervous system of a neurasthenic type and by disturbances of the blood count.

p-Toluidine

A liquid with b.p. 199 °C, of low toxicity, dangerous to the eyes. It has similar effects to aniline, causing cyanosis and irritation and inflammation of the bladder, connected with haematuria. It is less absorbed by the skin than aniline. The isomeric m-toluidine is more toxic than o-toluidine, while p-toluidine is more toxic than m-toluidine.

1-Naphthylamine

A grayish-white substance, m.p. 50 °C, toxic on ingestion. It mildly irritates the skin. It is not carcinogenic by itself, but since it always contains small amounts of the 2-isomer, it is labelled as a carcinogenic compound.

2-Naphthylamine

A white substance with m.p. 111 °C, slightly more toxic than the 1-isomer. It is carcinogenic, causing tumours in the bladder. The illness can occur after several years of working with this substance, and often even several years after work with the substance has ceased. The first symptom of the illness is the appearance of blood in the urine and difficulties with urination. Owing to its carcinogenic properties 2-naphthylamine is labelled as a poison and its manufacture is now almost discontinued worldwide.

o-Phenylenediamine (1,2-phenylenediamine)

A white substance, darkening in the air, with m.p. 103 to 104 °C. It is half as toxic as its p-isomer.

m-Phenylenediamine

A white substance with m.p. 63 °C. It is twice as toxic as the p-isomer and it possesses strong irritant properties.

p-Phenylenediamine (1,4-phenylenediamine)

A white substance darkening in the air, with m.p. 139 °C, considerably toxic. After absorption it affects the central nervous system, increasing blood pressure and causing inflammation of the mucous lining of the stomach. It also causes liver damage which may be lethal. It

irritates the skin, and if it enters the eyes it may cause permanent blindness. Two thirds of the people who come in contact with this substance are subject to sensitization, manifested by skin disease, conjunctivitis and the occurrence of bronchial asthma.

Phenylhydrazine

A liquid boiling at 243 °C (hydrochloride has m.p. 240 °C), very toxic, but causing cyanosis to a lesser extent than aniline. Phenylhydrazine damages the liver (jaundice) and kidneys. It affects the skin and causes various rashes. Phenylhydrazine is classified as a severe poison.

Pyridine

A colourless liquid with b.p. 115 °C. Mildly toxic on ingestion, toxic on inhalation. It is not dangerous to the skin, but very dangerous to the eyes. It has irritant properties, poisons the nervous system and it can damage the liver. Industrial poisonings are rather frequent. Acute poisoning is accompanied by vomiting and diarrhoea, reddening of the face, acceleration of the pulse and headaches. With high doses a high fever occurs, accompanied by delirium and the poisoning may end in death as a consequence of lung oedema. The symptoms mentioned already appear when 2 to 3 ml of pyridine are ingested. Chronic poisoning is accompanied by giddiness, insomnia, disturbed walking and epileptic convulsions. Paralysis of the facial nerves and deterioration of hearing have also been described. On repeated milder poisonings and in chronic poisoning the liver and kidneys are also damaged. Pyridine irritates the skin and causes increased sensitivity to light.

Cyanuric chloride

A white substance with m.p. 146 °C, not very toxic on ingestion. The fumes escaping from cyanuric chloride have irritant properties. It is very dangerous to the skin and eyes. According to some Soviet reports it is carcinogenic.

Morpholine

A colourless liquid with boiling point 128 °C. Low toxicity on inhalation or ingestion, but dangerous to the skin and very dangerous to the eyes. On strong inhalatory exposure the liver and kidneys are damaged. It is absorbed by skin and it is strongly irritant.

Urea

A white substance melting at 132.7 °C, harmless.

m-Nitrobenzenesulphonic acid

Not toxic on ingestion, very dangerous to the eyes and skin. Its sodium salt (Tiskan) is harmless and non-toxic.

Metanilic acid (3-aminobenzenesulphonic acid)

A white, non-toxic substance, not harmful to the skin or eyes.

p-Toluenesulphonic acid

A white substance with strong corrosive properties.

Optical brighteners based on 4,4′-diaminostilbene-2,2′-disulphonic acid

They do not irritate the skin, but they have photosensitizing properties. The non-sulphonated base, 4,4′-diaminostilbene, is carcinogenic.

6

Industrial Effluent Problems

The times are past when effluent from industrial plants could be drained directly into rivers or when waste gases could be expelled directly into the atmosphere without pretreatment. Today no industrial enterprise exists which is permitted to drain its waste waters into rivers, however large, or into the sea, without restriction, and the number of plants where waste gases are emitted freely into the air is constantly decreasing. The protection of water ways and the atmosphere has become a crucial problem in all industrially developed countries, principally because of the enormous development of industry and also the rapid increase of motoring, which in some places can be a definite danger to life. This health problem affects, of course, mainly the highly industrialized parts of the world, but — naturally — it also affects neighbouring areas. The purification of waste water has become a world-wide problem, which no country can solve alone. The USA, Europe and Japan are most affected. In Europe the German Federal Republic, Benelux and Great Britain, and of the socialist states, Czechoslovakia and the German Democratic Republic are the most affected. The expenditure of these countries on environmental protection is high and is increasing every year; it can amount to some 5 to 10 % (in some places even more) of total investments.

Contradictory demands on the one hand for an increase in production and on the other for the preservation of a healthy environment are influenced by ever stricter legal, economic and technical restrictions, which force the manufacturing sectors of the economy to increase production efficiency. This is achieved by using the results of technological development to increase productivity and simultaneously limit the noxious effects of the industry on the environment.

Optical brightening agents are produced in factories also used for the production of dyes. Production involves raw materials and intermediates of the same type as dyes, and therefore special measures for dealing with waste waters are not required. The effluent formed during

the production of optical brighteners is purified together with the waste waters from other dye production units in the same decontamination plants.

Waste waters from production plants where optical brighteners are actually applied (e.g. textile factories) do not require special purification equipment either; before the effluent is drained into rivers it is purified by the method customary for the respective application field. Thus a general survey of procedures used for the purification of waste waters from the production of dyes and dye intermediates will be presented here, and it will be assumed that such procedures are equally applicable to effluents from optical brightener plants.

6.1 The discharge of waste waters into rivers — some general considerations

The course of a river is divided into sections, each of which is labelled according to the intensity of pollution which must not be exceeded in that section. For example, if a certain pollution classification is assigned to section A of some river, then all users of water from this river, from its upper part down to section A, are forbidden to drain collectively more pollutants into the river than is allowed, nor more than any self-purification processes up to this point (or some artificial measure) are capable of eliminating.

In determining the extent of pollution of a water way, the content of insoluble inorganic substances, the smell, taste and coloration of the water are observed, together with the content of dissolved oxygen, the pH of water in the river and the content of toxic substances. Soluble inorganic salts are to all intents unaffected during the natural progress of the river. Their content is given by the degree of dilution, and it is usually determined by measuring the amount of chlorides present or by measuring the electrical conductivity of the water. Limiting values of smell and taste are not determined in waste waters as a whole, but only for specific substances that are known to be in the water. Thus, for example, hydrogen sulphide is detectable by human senses even at a 0.5 mg l^{-1} concentration, kerosene at a content of 1 to 2 mg l^{-1}, mineral oil at a content of 0.1 to 0.5 mg l^{-1}, chlorophenols in concentrations of 0.2 to 0.005 mg l^{-1}, etc. The determination of the limiting values of smell or taste in waste waters is not always sufficient, since secondary smells are created on mixing

various waste waters together or with the waters of the river, so that it is preferable to decrease the recommended values to one half or one third. The greatest danger in this respect is phenol contaminated water. Waste waters decrease the content of soluble oxygen in the river (the recipient), and the decrease of the oxygen content can be determined in several ways.

The water in a river stream should not have a pH lower than 6.5 and higher than 8.6. Calculation of the amount of acidic and basic waters that may be drained into the river is based on the familiar rules of hydro-chemistry. When acids and bases are drained into the river the pH of the water changes in consequence of the changes in the ratio of hydrogen carbonates present in the river water, and the free carbonic acid. The calculation of the respective permissible content of acids and alkalis in the recipient should be carried out on the basis of the lowest water-levels in the recipient river, and for maximum amounts of acids and alkalis drained by the enterprise.

Compound	Concentration (mg l^{-1})
ammonia	2
phenol	0.1
calcium hydroxide	20
chlorine	0.11
potassium chloride	5 000
magnesium chloride	1 000
sodium chloride	10 000
calcium chloride	3 000
chloride of lime	0.5
potassium cyanide	2
nitric acid	100
hydrochloric acid	50
acetic acid	50
sulphuric acid	50
sulphurous acid	0.5
oxalic acid	20
carbon dioxide	22.2
naphthalene	1.45
aluminium sulphate	1 000
cupric sulphate	0.1

The discharge of toxic substances into rivers is permitted only if there is no potential danger to the local population or to another industry using the water in which the waste waters were diluted, when the fauna and flora in the river are not damaged, or when complete elimination of toxic substances from the waste waters is impossible for technical and economic reasons. In the following survey concentrations of some toxic substances, noxious to fish, are given. For lower organisms this concentration is much lower.

6.2   Methods of purification of waste waters

Mixing industrial waste waters with sewage waters is considered the best method of purification of waste waters. This permits the construction of large purification plants with a good service and control under economically acceptable conditions. Before draining waters into these biological purification plants the industrial waste waters should be freed from toxic and foaming substances. Prepurification of waste waters is carried out by precipitation, chlorination and oxidation, extraction and adsorption. Where geological conditions permit, waste waters may be eliminated by pumping them to subterranean spaces.

When precipitation is used, the waste waters are saturated with various chemicals, such as lime, green vitriol or aluminium sulphate, thus converting the soluble components of the waste waters (for example sulphuric acid) into insoluble compounds. The precipitating agent should not be expensive. Sometimes it is possible to precipitate two types of waste waters by merely mixing them together.

The disadvantage of precipitation methods is that a considerable amount of sludge is formed that is very voluminous, poorly filtrable and sediments slowly. It must be collected from the sedimentation plants and transported to a storage place, where it accumulates.

The sludge is precipitated and coagulated in special purification stations or in a retention reservoir. The precipitation method is very important in the purification of waste waters and great attention has been paid to the improvement of the technique.

Oxidation and chlorination are usually mentioned together when the purification of waste waters is discussed, since chlorination with chlorine or hypochlorites is usually an oxidative process.

The purification of waste waters by oxidation of organic substances with hypochlorite may either lead to a stage when various water-insoluble substances are formed, which are thus mechanically separable, or to the last stage of oxidation, when the organic substance is oxidized to carbon dioxide. This latter situation is usually economically untenable and oxidation is usually taken to the first stage only.

The disadvantage of this method is that in addition to an oxidative effect, a chlorination effect is also operative to a certain extent, so that some chlorinated organic substances are formed often with a very disagreable smell (for example chlorophenols), which is especially disadvantageous when the river water is used for the preparation of potable water. Lately great attention has been devoted to the use of chlorine dioxide. It is much more effective than hypochlorite or perchlorate, and it possesses better oxidative and disinfecting properties than chlorine, its effect being practically independent of pH. Its disadvantage is the greater danger of poisoning and greater demands on service. The use of chlorine for the purification of waste waters can be advantageous only for factories which have their own means of producing chlorine.

Oxidation with air is used with advantage for the prepurification of sulphide waters. It is carried out in a column filled with pyrolusite as catalyst. The waste water is sprayed in the upper part of the column and air is introduced into the bottom part of the column. The consumption of air is 6 to 10 times larger than corresponds to theoretical calculation. The catalyst is regenerated directly in the column with dilute sulphuric acid. The oxidation process is very efficient but is dependent on the pH at which it is carried out. At pH 3 the oxidation take place so rapidly that the vent gases do not contain any traces of hydrogen sulphide or sulphur dioxide.

Waters which contain aromatic amines are oxidized in a similar manner using the same equipment. The amines are oxidized to quinones after only 30 minutes at pH 3. However, a considerable part is oxidized to tars, which form a deposit. After oxidation and settling the content of amines in waste waters is lowered by 80 to 90 %.

Extraction is used for the purification of waste water if the substance to be isolated occurs in sufficiently large amounts and if it can be recycled into the production process together with the extractant. This is the case, for example, in the elimination of aniline residues in waste waters. The water is extracted with nitrobenzene and both components are returned to the production cycle.

When only small quantities of contaminants are involved, preference is given to adsorption. Adsorption may be used both for regenerative and destructive purification of waste waters. The most suitable adsorbent is charcoal. A regenerative method has been elaborated for the purification of waste waters from the production of tar-dyes. It is based on adsorption on activated charcoal with subsequent desorption and regeneration of the sorbent. In this manner waste waters containing nitrobenzene, chloronitrobenzene, chlorodinitrobenzene, nitrotoluene, *o*-nitrophenol and *p*-nitrophenol, 2,4,5-trinitrophenol and other substances may be purified. In the destructive method of purification of waste waters, anthracite serves as adsorbent. Various techniques are used for this method. A modern method operates with a column in which anthracite is maintained in a fluid layer. It is introduced into the column continuously and after adsorption it is carried from it for regeneration, which is also carried out by the fluidization method. About 50 % of the anthracite is lost in this method. The results of the purification of the waste waters from the dye industry by this method are very good.

The greatest attention has been devoted to biological purification of waste waters. This method is applied mainly in city cleaning stations, but is also gradually being introduced for industrial waste waters, especially those from the production of dyes and intermediates. Special attention has been devoted to the study of the acclimatization of microorganisms. A number of chemical compounds present in the effluent from the dye industry are biologically degradable only after acclimatization of the microorganisms (which can take as much as several months).

In some countries experiments have been made to dispose of dye-industry waste waters by draining them into deep subterranean levels. However, this is realizable only where the geological composition of the soil permits.

In addition to the above methods of partial purification of waste waters the idea of the regeneration of water for its repeated use is becoming ever more important. This concept is based on the fact that some substances cannot be eliminated, even after repeated attempts at purification, so that they remain permanently in the rivers. This has deleterious consequences not only for the river flora and fauna, but especially when the river water is used for the production of drinking water. These substances, resistant to current processes of purification (so-called re-factors), are eliminated in special purification stations where various more complex methods are used in addition to the more usual purification methods, such as floating out on foam, purification on ion exchangers, etc.

On complete renovation, potable water can be obtained, or on partial renovation water is obtained which is suitable for agriculture and industry.

6.3   Exhalations

The part played by the chemical industry in contributing to air pollution by various gaseous emissions (among which the most important are sulphur dioxide, carbon monoxide, nitrogen oxides, chlorine, hydrogen chloride, ammonia and hydrogen sulphide) is less than 2 %. In dye-production plants the elimination of waste gases is generally solved by extracting them in adsorbers and water showers. In modern production plants (especially if they are located close to residential areas) each production unit has its own absorption system, the gases released from these systems are led into the central absorption system of the department and from there into the central absorbing system of the factory. The residual gases are then vented into a tall chimney.

7

Bibliography

1. K. Venkataraman: *The Chemistry of Synthetic Dyes*, Part V. Academic Press, New York, London 1971.
2. SVF — Fachorgan für Textilveredlung *19*, 404—487 (1964).
3. T. Rubel: *Optical Brighteners*. Technology and Applications, Noyes Data Corporation. Noyes Building, Park Ridge, New Jersey 07656, USA 1972.
4. J. Marhold: *Přehled průmyslové toxikologie* (*Survey of Industrial Toxicology*). Avicenum, Prague 1964.
5. G. Hommel: *Handbuch der gefährlichen Güter* (*Manual of Dangerous Goods*).Springer, Berlin, Heidelberg, New York 1973—1974.
6. J. Arient: *Přehled barviv* (*Survey of Dyes*). SNTL, Prague 1968.

Subject Index